THE MRICHCHHAKATIKA OF SUDRAKA

with Introduction, Critical Essays and a Photo-Essay

English Translation by

M.R. KALE

Edited by

KULJEET SINGH

MOTILAL BANARSIDASS PUBLISHERS
PRIVATE LIMITED • DELHI

in association with

WORLDVIEW PUBLICATIONS

4th Reprint : Delhi, 2019
First Edition : Delhi, 2015

Published by
MOTILAL BANARSIDASS PUBLISHERS (P) LIMITED
in association with
WORLDVIEW PUBLICATIONS
58 U.B. Bungalow Road, Delhi-110007

ISBN: 978-81-208-4010-2

Also available at
MOTILAL BANARSIDASS
41 U.A. Bungalow Road, Jawahar Nagar, Delhi 110 007
1 B, J S Compound, Kennedy Bridge, Nana Chowk, Mumbai 400 007
203 Royapettah High Road, Mylapore, Chennai 600 004
49, 8th F Main III Block, Jayanagar, Bengaluru 560 011
9, Panchanan Ghosh Lane, 1st Floor, Kolkata 700 009
Ashok Rajpath, Patna 800 004
Chowk, Varanasi 221 001

MLBD Cataloging-in-Publication Data
The Mrichchhakatika of Sudraka
with Introduction, Critical Essays, a Photo-Essay
English Translation by M.R. Kale, Edited by Kuljeet Singh
Includes Notes, Glossary and Bibliography
ISBN : 978-81-208-4010-2
I. Intoduction II. Critical Essays
III. Photo-Essay IV. M.R. Kale V. Kuljeet Singh

Printed in India
by RP Jain at NAB Printing Unit,
A-44, Naraina Industrial Area, Phase I, New Delhi–110028
and published by JP Jain for Motilal Banarsidass
41 U.A. Bungalow Road, Jawahar Nagar, Delhi-110007

To

Dr. BS Rattan, who taught me, *what*

and

Prof. TS Satyanath, who taught me, *how*

Yatho Hasta thatho Drishti.
Yatho Drishti thatho Manah.
Yatho Manah thatho Bhaava.
Yatho Bhaava thatho Rasa.

Where the hand (hasta) goes,
eyes (drishti) also should go there;
where the eyes go, mind (manah) also should go there;
where the mind goes, there the State (bhaava) should follow;
and where there is the State, there the Sentiment (rasa) arises.

Nandikeshvarà, *Abhinaya Darpanam.*
Translation by Manomohan Ghosh

Contents

Preface

This critical edition on Sudraka's *Mrichchhakatika* or *Mrichhakatika* or *Mrchhakatikam* (variously spelled) was initially aimed at fifth semester undergraduate students of English Literature amidst the mayhem of recurrent changes in the courses of the University of Delhi. While I was planning to create a grid of contents for this critical edition, the guidelines for the newly implemented CBCS (Choice Based Credit System) were announced and this text was also prescribed for first semester English Honors students, which I think is a formidable one for the students coming from a school system like ours. Creating an edition which may suit the creative and intellectual faculties of the third year FYUP students and responding to the simplistic inquisitiveness of the first year students under one cover, remained the most arduous task. The acute sense to construct a text which can inform, enthuse and inspire led to multiple constructive changes, which I think will help the students holistically.

With a lot of syntactical concerns and readability factors in the available translation, I negotiated gently with the former and focussed majorly on the latter, trying to make the text more readable. I have removed parenthesis from it, wherever required, thus making it (as it looks to me now) lucid and comprehensible and provided edited notes from the annotated text by MR Kale, along with a few additional responses. In order to reach closer to the original while not being adept in Sanskrit, I have taken guidance from the available Hindi translations by Mohan Rakesh and Rangey Raghav. Having said this, the translation in English will not, in any way, provide the linguistic joy of the original in which Sanskrit and seven dialects of Prakrit were used, needless to mention the musicality and poetic quality of the play. Moreover, the fun of comic elements the character of Sakar brings through lisping of "sh" in place of "s" could not be achieved in the English translation.

Though this edition underwent a long and detailed creative

process, I would be glad to receive feedback from students and teachers to fill in its gaps, if any remain. For this volume, I am indebted to Prof. TS Satyanath for inspiration and suggestions; Prof. Indira Kaul for providing material and promising to contribute an essay in the revised edition; Prof. Devendra Raj Ankur for consenting to the use of his exhaustive essay; Saheb Kaur for translating the same from Hindi; Prof. KS Rajendran for providing photographs; National School of Drama for allowing us to use them; Simranjot for reading out the essays; Kanav Gupta for his critical responses; Menka, Shikha and Kanupriya for proof-reading it thoroughly. I am indebted to Motilal Banarasidass and Sachin Rastogi for publishing it.

KULJEET SINGH

Publishers' Acknowledgement

The publishers wish to thank the following for the use of material:

Farley P. Richmond, for the chapters, 'Origins of Sanskrit Theatre' and 'Characteristics of Sanskrit Theatre and Drama' from the book Farley P. Richmond, Darius L. Swann, and Phillip B. Zarrilli, eds. *Indian Theatre: Traditions of Performance*. Motilal Banarsidass, New Delhi. 2007.

Madhusudan Pati, for the essay, '*Mrichchakatikam*: Thematic and Tonal Patterns' from his book *Sanskrit Drama: Essays in Revaluation,* Amar Prakashan, Delhi, 1991.

Devendra Raj Ankur, for the essay, '*Mrichchhakatika* : A Reading', *Sangana* Jan-Mar 2012, pp. 62-85, translated for this edition and Interview extracted from the brochure of Shudrak's *Mrichchhakatika,* National School of Drama, November 19-22, 1993.

Every effort has been made to trace the owners of copyrighted material, but if any have been inadvertently overlooked, the publishers will be pleased to make necessary arrangement at the first opportunity.

Introduction

Kuljeet Singh

Classical Drama: Indian and Greek

The following *sloka* (verse) given in chapter one of *Natyashastra* precisely articulates the thematic range of the classical Indian drama in the *Pancham Veda,*[1]

> There is no wise maxim, no learning, no art or craft, no device, no action, that is not found in the drama (*natya*).
>
> – Bharata, *Natyashastra*, trans. Manomohan Ghosh

In the same chapter, readers are told about the mythical account of the origins of *Natyashastra* and its religious roots.[2] *Natyashastra*, loosely translated as 'a compendium of theatre', is the single most important source to decode and understand the character of classical Indian theatre, poetics, aesthetics, dance and music. An encyclopaedic compendium, it also contains elements of architecture, painting, prosody, language, grammar, phonetics and other aspects of theatre and draws on disciplines as diverse as philosophy, psychology, mythology, ritual and geography. It is interesting to note here that gods were concerned about the waning of *dharma*[3] in human beings with each *yuga*,[4] leading to suffering. Under the aegis of Indra they requested Brahma to create a *Kridaniyaka*[5] which would be capable of providing both *drishya*[6] and *shravya*[7] to all *varnas*[8]. Upon completion of this theatre theory — *Natyashastra* — Brahma encouraged gods to receive this distinctive creation and use it to dispense the meaning of Vedas to humanity to salvage it from pain and sorrow. What is intriguing is that the gods refused the offer. Abhinavagupta,[9] the most reliable commentator of *Natyashastra*, clarifies that the reason the gods did so was because they only knew happiness, whereas *natya* is the saga of happiness as well as suffering.

Although its historical dating is disputed, classical Indian drama seems to have developed and emerged as a sophisticated form of public literature in the city of Ujjain, in Northern India, during the Gupta period (circa 390-470). The prologues of plays attributed to the Buddhist poet Asvaghosa, who is believed to have lived at the court of Kushan ruler Kanishka in the first or second century CE, verify this. Plays of Kalidasa, Sudraka and Vishakhadatta are acknowledged in reference to the genre of early monologue plays and popular theatre.

Inadequately equated with "acting", *abhinaya* literally means, "carrying towards", which more comprehensively explains how the actor (in Sanskrit, *nayaka*) transacts the meaning of drama towards the audience, or the *prekshaka* (spectator, literally, "the one who sees"). Two inferences can be drawn from this: one, *Natyashastra* or classical Indian theatre greatly emphasizes the visual aspect of performance, as seen in the concepts of *drishya* and *shravya*, and the literal definition of *prekshaka*. Second, we get the core idea of *rasa*[10] meaning sentiment, or better framed, the "transaction of sentiment on stage". While the former contrasts Sanskrit drama to the then existing lyric and epic forms (whose composition takes little or no notice of the audience) and brings visual poetry to the domain of public literature, the latter makes a tangible distinction from its Western counterpart: on Indian stage we *transact* sentiments whereas Western theory talks of *feeling* the sentiment.[11] What the audience experiences while watching a performance is a *taste* of the sentiment and not the *real* sentiment. Thus one can suitably argue that the focus in Indian dramaturgy is on the "skill of acting" rather than "capacity to feel", as *Natyashastra* also emphasizes the "physical manifestation to the sentiment rather than becoming sentiment itself . This idea is further strengthened in the four components[12] of *abhinaya*, which lay emphasis on the accuracy of expression upon the feeling of emotion. While acting styles belonging to the East are accused of being "sentimental", the Western acting style, which is by definition emotionally driven,[13] looks emotionally subdued. Indian theory does not reject *sentiment*, it only upholds *technique*. Greek dramaturgy, on the other hand relies completely upon the *emotional purging* of the character, which is shared by the actor and the character.

An American Indologist,- Barbara Stoler Miller[14] comments,

> Like Aristotle, Bharata stresses emotional satisfaction through action. Although their modes of ordering experience are significantly different, Greek and Indian dramas were conceived and performed as sources of pleasure and insight for the audience. These two undertakings of drama crystallize our perceptions of classical Greek and classical Indian culture.

With regard to the two diametrically opposing worldviews, she further explains,

> Greek tragedies present impossible heroic struggles against established orders in which the fate of every person in the story and, by extension, the whole human audience is implicated. When Aristotle discusses the emotional effect of a drama on an audience, he speaks both of sympathy and the excitement through which the play may bring about catharsis, that purgation of emotions which restores a balance of conflicting forces. In Greek tragedy the conflict is between the actors in the drama and a superior force, whether society or destiny. It involves a catastrophe that arouses fear and pity in such a way that these emotions are purged, and order is restored in the soul of the audience. Indian dramas represent emotions in the theatrical universe of symbolically charged characters and events in order to lead the audience into a state of extraordinary pleasure and insight. The goal of Sanskrit drama is to reestablish emotional harmony in the microcosm of the audience by exploring the deeper relations that bind apparent conflicts of existence. The manifestation of these relations produces the intense aesthetic experience called rasa.

This discussion will not go into the details of Sanskrit drama as the reader would be acquainted with substantial information on its origins and characteristics in the essays that follow. Instead, this section will conclude with a discussion of the idea of *prekshaka* as *sahridaya.*[15]

The first chapter of *Natyashastra* claims various reasons for its genesis: entertainment, edification of all the *varnas*, and bringing humanity back to the path of *dharma* in the age of ignorance and suffering. In Chapter 27, Bharata[16] outlines the characteristics of an ideal spectator (*Prekshaka*) of a performance:

> Those who are possessed of [good] character, high birth, quiet behaviour and learning, are desirous of fame, virtue, are impartial, advanced in age, proficient in drama in all its six limbs, alert, honest, unaffected by passion, expert in playing the four kinds of musical instrument, very virtuous, acquainted with the costumes and Make-up, the rules of dialects, the four kinds of Histrionic Representation, grammar, prosody, and various [other] Sastras, are experts indifferent arts and crafts, and have fine sense of the Sentiments and the Psychological States, should be made spectators in

witnessing a drama" (NS- XXVII.50-53)

Who are these people of "high birth and proficient in various Sastras"? The egalitarian idea that the *Pancham Veda* is meant for the consumption of all *varnas* from Brahmins to Shudras, as expressed in the first chapter of *Natyashastra,* seems to be contradicted here and needs further enquiry, as no substantial critical responses have been made to the same.

Bhasa and Sudraka

The dating of Sanskrit dramatists is a perennial debate in the history of Indian literary history, but most scholars[17] and historians[18] agree that Bhasa was the first Sanskrit poet and the first Indian playwright followed by Kalidasa[19] and Sudraka. His plays were discovered by chance in the early years of the twentieth century, among the performance scripts used in the Koodiyattam tradition of Kerala.[20] Ganapati Shastri, a local scholar, attributed thirteen of these texts to Bhasa and published them between 1912 and 1915, though their authorship has been contested. His most popular, *Swapnavasavadatta* (Vasavadatta in the Dream) deals with the popular tale of King Udayana and his queen Vasavadatta, *Urubhanga* (Breaking of the Thigh) a one-act play, departs from the forbidden imperatives of *Natyashastra* by demonstrating the death of Duryodhana on stage. Though the plot belongs to *Mahabharata,* Bhasa made major alterations in *Urubhanga,* the most extreme of which is his portrayal of Duryodhana — whereas in the *Mahabharata,* he is portrayed as a villain, in *Urubhanga* Bhasa gives him more human qualities. While tragedy is rare in Sanskrit drama, Bhasa's presentation of Duryodhana's side of the tale adds a certain tragic element to the play.

Most of the plays attributed to Bhasa are inspired by the *Ramayana* or *Mahabharata.*[21] While the seven-act *Pratimanataka* (Statue Drama) and the six-act *Abhishekanataka* (Coronation Drama) deal with *ramakatha* (tale of Rama), Mahabharata is the source of six short plays: *Pancharatra* (Five Nights), in which Drona is asked to bring news of the incognito Pandavas within five days; *Karnabhara* (Karna's Burden), on Karna's tradegy; *Dutavakya* (The Messenger's Word), where Krishna goes to Duryodhana as the Pandavas' messenger for a peaceful agreement to avoid war; *Duta-Ghatotkaca* (Ghatotkacha as Messenger), where

Ghatotkacha delivers Krishna's counsel and final warning to Duryodhana; *Madhyamvyayoga* (Vyayoga of the Middle One), about Bhima, the middle Pandava brother and Urubhanga on the death of Duryodhana, as mentioned before. The remaining plays are longer, based on legendary episodes. Bhasa's plays form a richly varied corpus both thematically and generically, comprising both full length *natakas* as well as one act plays. Their rich stage potential, in terms of ingenious employment of theatrical techniques, remarkable presence of dramatic irony and humour, powerful representations of emotions, and vast area of human experience, have led to regular productions by modern Indian directors.

Daridra Carudatta (The Poor Charudatta) is a fragment in four acts without the initial and the final verses. This play proves to be the *bija* (germ) of Sudraka's *Mrichchhakatika*, a *prakarana* (a play in ten acts).[22] According to common belief, Sudraka was an Indian king and playwright. Three plays are credited to him: *Mrichcchakatika* (*The Little Clay Cart*), *Vinavasavadatta*, and *Padmaprabhritaka*, a *bhana* (a short one-act monologue). Although there is no credible evidence, Sudraka has been identified as the *nom de plume* of a third century Abhira king, Sivadatta. Abhiras were herdsmen, and it is likely that the character of Aryaka (in *Mrichchhakatika*) may be one, as his father is called Gopala, which means cowherd. Sudraka has also been identified as the founder of Andhrabhritya dynasty (second century BC), as Vikramaditya, famous for Vikrama era (fifty six BC) and as the Sakasatrap Rudra or Rudradaman (one hundred thirty AD).

Set in the ancient city of Ujjayini, Sudraka's *Mrichchhakatika* begins with a love story of a poor yet virtuous Brahmin-merchant, Charudatta and a *nagarvadhu* (coutesan), Vasantasena. Sakara, also referred to as Samsthaṇaka, the brother-in-law of the corrupt King Palaka, is the comic-villain of the play, and is completely in love with Vasantasena and chases her aggressively. In the third act, Sharvilaka, the thief, breaks into the house of Charudatta to steal valuables in order to ransom the freedom of Madanika, who is the maid-servant of Vasantasena. He manages to steal the gold given to Charudatta by Vasantasena as a deposit. Despite the odds in the first five acts of the play, love blossoms between Charudatta and Vasantasena, beautifully depicted in the fifth act. The subplot of the play begins with the sixth act and runs parallel till the end. The play reaches its climax with the accidental swapping of carts by Aryaka, a rebel who has escaped the prison to usurp the position of the corrupt and despotic King Palaka, and Vasantasena, who is going to meet her lover Charudatta in the *Pushpakarandaka* garden, where Sakara is also waiting for his cart.

Hailed as a Shakespearean play,[23] written a few centuries before Shakespeare, Sudraka's *Mrichchhakatika* is one of the most significant Sanskrit dramas for more than a few reasons. Set in the ancient city of Ujjayani, *Mrichchhakatika* is "rife with romance, comedy, intrigue and a political subplot detailing the overthrow of the city's despotic ruler by a shepherd, [and] is notable among extant Sanskrit dramas for its focus on a fictional scenario rather than on a classical tale or legend. *Mrichchhakatika* also departs from the traditions enumerated in the *Natya Shastra* that specify that dramas should focus on the lives of the nobility and instead incorporates a large number of middle and lower-caste characters who speak a wide range of Prakrit dialects."[24]

Much celebrated in the West because of its plot structure, following several successful nineteenth century translations and stage adaptations,[25] *Mrichchhakatika* remained the prominent drama in Sanskrit, widely translated, adapted and performed internationally.

Mapping *Mrichchhakatika*

After Independence, the first stage production of *Mrichchhakatika* was directed by Habib Tanvir[26] in 1958. Adapted as 'Mitti Ki Gadi' by Begum Qudsia Zaidi[27] for Hindustani Theatre[28], it is considered a masterpiece for its innovative idiom and primitive flavor. In the interview published in Enact[29], March 1974, Tanvir underlines the experiment of having folk and urban actors in the same space for the same play, with regard to his version of *Mrichchhakatika*. He adapted the Sanskrit *slokas* (verses) 'in folk tunes and meters, guided by the belief that some of these tribal and semi-tribal folk tunes are so primitive that by introducing these tunes a certain primitive flavor would be given to play'. A Bharatanatyam dancer was cast to play the courtesan Vasantasena and the folk actors played some of the important plebeian roles, 'accompanied by mask, make-up and highly stylized costumes'. This must have lent to the production a hybrid of folk style and the modern idiom.

In the last three decades, various productions of *Mrichchhakatika* had been staged in Delhi, based on the translation by Mohan Rakesh[30], the eminent Hindi playwright, entitled 'Mitti Ki Gaadi'. Three of the theatre productions were produced by National School of Drama (NSD), Prof. Devendra Raj Ankur directing it with the second year students of NSD in 1992, followed by Prof. Robin Das in 1998 and most recently, Prof K.S. Rajendran in 2008. While Devendra Raj Ankur focussed 'on find(ing) meaning in the dramas of our own tradition as Sanskrit Theatre is an ensemble of alienation, connection, speech, story-telling *et al*, which are considered to be essential elements for an actor today', Robin Das's stage design remained unsurpassable and K. S. Rajedran's production laid greater emphasis on the erotic (Sringara).[31] Abhivyakti, the theatre society of Miranda House, University of Delhi has also produced *Mrichchhakatika* in Hindi in 2009 as their annual production, which was directed by an NSD alumnus, Ramji Bali. However, no major Indian theatre director has yet attempted to stage Sudraka's *Mrichchhakatika* from the feminist perspective with a *ganika* (courtesan) as the *nayika* (heroine) of the play either in classical or contemporary idiom.

The film *Utsav* (1985) by Girish Karnad was based on *Mrichchhakatika*, with Shekhar Suman playing Charudatta and veteran actor Rekha playing Vasantasena, along with Amjad Khan as Sutradhar, Shashi Kapoor as Samsthanaka and Shankar Nag enacting Sarvilaka.

This volume includes an annotated text translated by M.R. Kale along with relevant background essays and criticism, to enable students of literature to understand the long tradition of theatre in India on one hand; and on the other, its subsequent negotiations with the West, traversing, accommodating, negotiating and becoming what we call Modern Indian Drama today. It also includes an interview with Prof. D.R. Ankur which briefly articulates issues concerning his production, and a photo-essay with the photographs of the production by Prof. K.S. Rajendran.

The first two essays, 'Origins of Sanskrit Drama' and 'Characteristics of Sanskrit Theatre and Drama' by Farley P. Richmond are foundational essays which deal with the mythological origin and the theoretical framework of *Natyashastra*, respectively.

The annotated text is followed by a glossary of Sanskrit terms and key concepts which should be of immense help to students who have been initiated into reading a classical Indian dramatic text. The glossary is created considering the requirements of the first-timers and it is by no means exhaustive and complete. Seekers may look into the bibliography for further details.

Madhusudan Pati's nuanced essay 'Thematic and Tonal Patterns' problematises the idea of 'art and trade' in the context of *Mrichchhakatika.* It eventually embarks on a discussion of realism on Sanskrit stage and situates Sudraka's 'civilized art' in the significant cusp between the ancient and medieval dramatic traditions.

Prof. Devendra Raj Ankur's critical essay, '*Mrichchhakatika*: A Reading', translated from Hindi by Saheb Kaur, discusses Bhasa's *Daridra Charudatta* (Poor Charudatta) as the source text of the play in detail, paying tribute to the imaginative ingenuity of Sudraka in creating a trenchant *Prakarana* with a twin plot from a four act play, letting both the plays interact with and nourish each other. Furthermore, he deliberates on the unique dramatic strategies — ornaments and carts — in *Mrichchhakatika*, comparing it with myriad dramatic texts across time and space. Designating *Mrichchhakatika* as the first Indian political play, he concludes by analyzing the "powerful and lively political, social and religious circumstantial" narrative that effortlessly establishes *Mrichchhakatika* as an "incredibly contemporary" classic.

The last essay in the edition, 'A Photo-essay on *Mrichchhakatika*' by the editor, uses photographs of the 2008 production by Prof. K. S. Rajendran to construct a photo narrative. The objective behind this essay is to facilitate at least the *drishya* (visual) to students of literature in the classroom, if the *shravya* (audible) is not practically possible.

Works Cited

Keith, A. Barriedale. (1924). *The Sanskrit Drama.* London: Oxford University Press. 1971.

Lal, Anand (2004). ed. *The Oxford Companion to Indian Theatre.* New Delhi: Oxford University Press. 2004.

Miller, Barbara Stoler. (1994). *Masterworks of Asian literature in Comparative Perspective : A Guide for Teaching.* NY: M. E. Sharpe, 1994.

Prasanna. (2013). *Indian Method in Acting.* National School of Drama. 2014.

Richmond, Farley P. (1990). Farley P. Richmond, Darius L. Swann, and Phillip B. Zarrilli, eds. *Indian Theatre: Traditions of Performance*. Motilal Banarsidass, New Delhi. 2007.

Sastri, Gaurinath (1960). *A Concise History of Classical Sanskrit Literature*. Calcutta: Oxford University Press. 1960.

Notes

1. *Natyashastra* is been given the status of the Fifth Veda (*Pancham Veda*) which is believed to have derived its form from the four existing Vedas. It consists of four elements: *pathya* or text, the art of recitation and rendition in performance taken from the *Rig Veda*; the second is *gita* or songs, including instrumental music from the *Sama Veda*; *abhinaya*, inadequately translated as "acting", the technique of expressing the poetic meaning of the text and communicating it to the spectator from the *Yajur Veda*, and fourth is *rasa* or aesthetic experience from the *Atharva Veda*.
2. This is further elaborated in the chapter *Origins of Sanskrit Theatre* by Farley P. Richmond.
3. *Dharma* is a derivation from the root *dhra*, which means 'to hold, maintain, keep' and takes a meaning of 'what is established or firm', and hence 'law'.
4. Hindu worldview conceptualizes time in four *yugas*, an epoch within a four age cycle. A complete *Yuga* starts with the *Satya Yuga* (golden age; time of truth and *dharma*), followed by *Treta Yuga* (the dharma bull, which symbolises morality, stood on three legs during this period) via *Dvapara Yuga* (the dharma bull stood on two legs, emphasizing degeneration of values *et al*) into a *Kali Yuga* (dark ages, time of darkness and non virtue). It is believed that our present time is a *Kali Yuga*, which started with the end of the Mahabharata war.
5. *Krida* literally, 'to play'. *Kridaniyaka* literally, 'a toy'.
6. Pertaining to the visual.
7. Pertaining to the auditory.
8. The quadruple classification in ancient Hindu literature of all humankind, and all created beings, in principle, into four *varnas*: *Brahmins*: priests, teachers and preachers; *Kshatriyas*: kings, warriors and soldiers; *Vaishyas*: traders and merchants and *Shudras*: labourers and peasants.
9. One of India's important philosophers, mystics and aestheticians, famous for *Abhinavabharati*, a commentary on Bharata Muni's *Natyashastra*.
10. This concept is briefly elaborated in *Characteristics of Sanskrit Theatre and*

Drama by Farley P. Richmond. For a detailed account, read the prescribed background prose reading: Bharata, *Natyashastra*, tr. Manomohan Ghosh, Chapter VI: Sentiments. Also see Glossary.

11. Prasanna. *Indian Method in Acting.* Chapter 41, pp 180-81.
12. See Glossary.
13. The idea of catharsis in Greek dramaturgy. Catharsis is a participatory and emotionally driven process.
14. Miller, Barbara Stoler. "Two classical Indian plays: Kalidasa's *Sakuntala* and Sudraka's *Little Clay Cart.*"
15. Connoisseur, or literally, "with heart".
16. Bharata, *Natyashastra*, trans. by Manomohan Ghosh.
17. Adya Rangacharya, *The Indian Theatre* & Farley P. Richmond, *Indian Theatre: Traditions of Performance.*
18. Romila Thapar, *Sakuntala: Texts, Readings, Histories*, Barbara Stoller Miller, *Theatre of Memory: The Plays of Kalidasa* & AB Keith, *The Sanskrit Drama.*
19. See Glossary.
20. The oldest dance drama form in India, still extant in Kerala.
21. Ananda Lal, *The Oxford Companion to Indian Theatre.*
22. A detailed analysis of Bhasa's *Charudatta* and Sudraka's *Mrichchhakatika* can be found in the essay by Devendra Raj Ankur in this volume.
23. Rajiv Velicheti in his brief essay 'Of the Unique Phenomenon Mrichchhakatika' included in the brochure of the NSD production of *Mrichchhakatika* in 1998 contests the idea of Sanskrit drama as 'too unworldly, too sweet, too poetic'.
24. Richmond, Farley P. (1990). Farley P. Richmond, Darius L. Swann, and Phillip B. Zarrilli, ed. "Characteristics of Sanskrit Theatre and Drama" in *Indian Theatre: Traditions of Performance.* Honolulu: University of Hawaii Press. pp 55-62.
25. Most notable ones are Gerard de Nerval and Joseph Mery's highly romanticized French adaptation titled *Le Chariot d'enfant* that premiered in Paris in 1850, as well as a critically acclaimed interpretation by Victor Barrucand called *Le Chariot de terre cuite* that was produced by the Theatre de lceuvre in 1895. The play was translated into English by Arthur W. Ryder in 1905 as *The Little Clay Cart.* It had previously been translated as *The Toy Cart* by Horace Hayman Wilson in 1826. Ryder's version was enacted at the Hearst Greek Theatre in Berkeley in 1907 and in New York in 1924 at the Neighborhood Playhouse, which was then an off-

Broadway theatre; at the Theater de Lys in 1953 and at the Potboiler Art Theater in Los Angeles in 1926, when it featured actors such as James A. Marcus, Symona Boniface and Gale Gordon.

26. Indian theatre director (1923-2009), renowned for his work in Indian folk styles – 'Nacha' and 'Karma' from Chhattisgarh and Madhya Pradesh. Founder of theatre organization, Naya Theatre (1959), he adapted Shakespeare's *Midsummer Night's Dream* in folk idiom as 'Kaamdev Ka Sapna Basantritu Ka Apna' as well along with many other significant works.
27. Writer, social worker, and theatre practitioner, she is best remembered for her pioneering role in establishing the Hindustani Theatre in Delhi in the 1950. She died in 1960.
28. In 1954 Begum Qudsia Begum met Habib Tanvir, who is today remembered as the founder of Naya Theatre and together, they established the Hindustani Theatre.
29. Theatre Journal by Rajinder Paul, which became very popular in the seventies.
30. Hindi playwright, known for his play 'Aadhe Adhure', considered as a modern classic.
31. http://www.thehindu.com/todays-paper/tp-features/tp-sundaymagazine/an-ancient-world-comes-alive/article1436915.ece

Sanskrit Theatre

Origins of Sanskrit Theatre

Farley P. Richmond

The traditional legend relating to the origin of drama and theatre serves as a reminder that the ancient Indians were fond of attributing the creation of all important institutions to divine ingenuity. The tale is found in Chapter I of the *Nāṭyaśāstra,* attributed to Bharata (Ghosh 1961 and 1967), and runs as follows:

One day wise sages approached Bharata, an elderly priest, during an interval in his studies and asked him to relate the story of the origin of drama. Bharata commanded them to get themselves "cleansed, be attentive, and hear" (Ghosh 1967, 2). At a time when people were addicted to sensual pleasures, desire and greed, and jealousy and anger, the god Indra along with some of the other gods approached Brahmā, the creator, and requested that he create an object of diversion which would be audible as well as visible. Indra asked that all members of the social order be permitted to hear it: the Brahmans *(brāhmaṇa),* members of the priest caste; the Kshatriyas *(kṣatriya),* members of the warrior caste; the Vaisyas *(vaiśya),* members of the trades caste; and the Sudras *(śūdra),* members of the peasant caste. Brahmā fell into deep meditation and thought on the four books of hymns sacred to the Hindus. He concluded that he should make a fifth Veda and call it drama *(nāṭya).* To do this he took recitation from the *Ṛg Veda,* the oldest book of hymns, song from the *Sāma Veda,* acting from the *Yajur Veda,* and aesthetics from the *Atharva Veda.*

Then Brahmā asked Indra to compose plays based on semihistorical tales and have them acted by the gods. Indra modestly refused on the grounds that it was not proper for the gods to act in plays. Instead he proposed that Brahmā request the priests "who know the mystery of the Vedas and have fulfilled their vows, to maintain and practice the new creation" (Ghosh 1967, 5). Whereupon Brahmā summoned Bharata and asked him and his hundred sons to learn the dramatic art from him.

Bharata agreed. The sons were assigned different roles and Bharata set about preparing the first performance. To lend grace to the occasion Brahmā created nymphs to assist in the acting and dancing. Musicians were employed to play and to sing. For the first performance Brahmā chose the occasion of the celebration of Indra's banner pole, which concerned the victory of the gods over the demons. Brahmā and the other deities were so pleased that they gave gifts to the actors as a token of appreciation for undertaking the performance.

But when the show got underway, the demons took offense and caused the actors to forget their lines and physical movements. When Brahmā ascertained that malevolent spirits were responsible for the disturbance, Indra took his banner pole and smashed them to pieces. The gods rejoiced and suggested that henceforth the banner pole should be used as a symbol of the divine protection afforded actors. The performance proceeded but once more the demons attempted to interrupt the show. Bharata requested Brahmā to take some action which would assure the successful completion of the performance.

Brahmā requested Viśvakarmā, the heavenly architect, to construct a theatre structure which would be secure from further disruptions and which would ward off evil spirits. When the building was completed, Brahmā assigned various gods the task of protecting different parts of the structure. Brahmā himself undertook to sanctify and preside over the center of the stage.

In a mood of reconciliation, the gods asked Brahmā to pacify the malevolent spirits. The demons came forward to express their grievances. They contended that the play depicted them in an unfavorable light and showed favoritism toward the gods. In replying to their complaints Brahmā articulated the objective of drama and theatre. He said, "The drama will thus be instructive to all, through actions and states depicted in it, and through sentiments, arising out of it" (Ghosh 1967, 15). Drama is not meant to represent one individual or class of individuals exclusively. "It will [also] give relief to unlucky persons who are afflicted with sorrow and grief or [over]-work, and will be conducive to observance of duty as well as fame, long life, intellect and general good, and will educate people. There is no wise maxim, no learning, no art or craft, no device, no action that is not found in drama" (Ghosh 1967, 15).

In conclusion Brahmā charged the gods to perform a sacred ceremony *(yajña)* in the playhouse and the actors likewise to offer a sacrifice *(pūjā)* to the presiding deity of the stage "which is similar to the (Vedic) sacrifice." Brahmā warned that those who did not offer the sacrifice would be "born as an animal of lower order," and of those who observed the rules and practices of sacrifice, he predicted that they "will attain auspicious wealth and will [in the end] go to heaven" (Ghosh 1967, 17).

This simple tale gives us a wealth of useful information concerning the way the theatre event is regarded in the *Nāṭyaśāstra*. It is obvious that theatre is special, set apart from daily life. It is composed out of material which is considered sacred (namely, the Vedas). It is to be performed by specialists, whose hereditary line may be traced to Bharata himself, a member of the highest social order, the priestly caste. Theatre requires special knowledge above and beyond that required for the performance of other ritual events, including sacrifices. Brahmā makes it plain that the actors were to undergo extensive training before presuming to perform. This training originally came from the creator himself and was passed on to man through Bharata and his hundred sons. It required the development of special skills in dance, music, recitation, and rituals. The theatre fulfilled numerous purposes, not the least of which was to educate and entertain. Like other important ritual events, it was to be performed according to a prescribed order and served as a sacrifice in honor of the gods.

Historical Evidence

The traditional legend offers useful insights into the origins of theatre; history is not quite so generous. From the scant evidence that remains, it is impossible to fix an exact date for the creation of theatre in India. Yet there is good reason to believe it came into being sometime between 200 and 100 B.C. By this period all the elements were present which could have given rise to it. There was a lively tradition of dance, music, and song, all of which were already old and respected branches of the performing arts. Plus there was a rich heritage of myths and legends, historical and semihistorical tales. Also communities of specialists in storytelling and the performing arts may be traced to this and earlier periods. To whom the idea may be attributed and what inspired it are lost to us. We know that the oldest fragments of Sanskrit plays which date from the first century A.D. reveal a dramatic structure already highly developed. The patterns that are evident in these ancient works were to be refined and repeated for nearly twelve hundred years. In the following pages I will explore some of the evidence that leads to the conclusion that Indian theatre originated during the second century B.C.

Early Visual Clues

Indian culture is very old. The Indus Valley Civilization developed a city culture along the Indus river basin in northwestern India about 2300-1750 B.C. Although excavations undertaken at the ancient cities of Mohenjodaro and Harappa reveal a wealth of detail about the early inhabitants, there is nothing to suggest that they knew theatre. However, there are hints of the existence of dance. Small copper and stone figurines of what are obviously nude females have been catalogued in the archaeological record as dancers. If it is true that the figurines depict dancers, music and song may also have been part of the rituals and festivals of the peoples of this civilization.

Sacred Literature

About 1500 B.C. the Indus Valley Civilization came to an abrupt and unexplained halt. Shortly thereafter the country was visited by successive waves of people from the West, people riding on horseback. Their culture initially may have been inferior to that which existed along the Indus river, but it was destined to become the dominant civilization of the region. The new peoples were later called Aryans and a record of their religion and philosophy of life is found in collections of sacred hymns dedicated to supernatural forces. The earliest hymns are found in a collection known as the *Ṛg Veda*. Although an accurate date for the composition of the *Ṛg Veda as* still to be established, scholars generally agree it was composed between 1500 and 1000 B.C. The hymns yield an interesting and perhaps misleading set of clues regarding the formation of theatre. Among the hymns are a few which are short dialogues between two and sometimes three individuals. The following are the concluding verses of one such hymn. The encounter concerns Yami who attempts, without success, to seduce her twin brother Yama.

Yama: Sure there will come succeeding times
When brothers and sisters will do acts unmeet for kinsfolk.
Not me, O fair one,—Seek another husband, and make thine arm a pillow for thy consort.

Yami: Is he a brother when no lord is left her?
Is she a sister when destruction cometh?
Forced by my love these many words I utter.
Come near, and hold me in thy close embraces.

Yama: I will not fold mine arms about thy body:
They call it sin when one comes near his sister.
Not me,—prepare thy pleasures with another.
Thy brother seeks not this from thee, O fair one.

Yami: Alas! thou are indeed a weakling, Yama;
We find in thee no trace of heart or spirit.
As round the tree the woodbine clings,
Another will cling about thee girt as with a girdle.

Yama: Embrace another, Yami;
Let another, even as the woodbine rings the tree, enfold thee.
Win thou his heart and let him win thy fancy,
And he shall form with thee a blest alliance.

(Griffith 1963, 391-394)

Interesting though it is to speculate that a short exchange such as this might have given rise to the development of full-scale drama, several factors argue against it having done so. A. B. Keith has pointed out that dialogues are unique to the *Ṛg Veda* and died out in the later Vedic literature (1924, 15). Thus it was apparently not popular at this period in time. Also we are unclear as to how the hymns were recited or sung. Possibly two priests impersonated the different roles but this, too, is sheer speculation.

Dramatic Rituals

During the Vedic age (c. 2000-1500 B.C.) ritual practices arose which are highly dramatic in character. In general Vedic rituals include special hymns which are sung or recited in praise of various elemental powers and deities. The ceremonies are complex and include elaborate rites which incorporate symbolic gestures and physical actions often accompanied by percussion and music. In order to carry out the ceremonies, the priest-performers separate themselves from the profane world by rituals, and in so doing, they often take on the role of a personality other than their own. For example, in the death ceremonies conducted in Kerala today by certain sects of Tamil Brahmans, a priest takes the role of a departed family member and converses with the living relatives who have commissioned the rite.

Conflict is the center of some of the rites. The Mahāvrata rite celebrates the strengthening of the sun at the winter's solstice, a contest between Vaiśya (white) and Śūdra (black). Besides bearing the obvious interpretation of the victory of light over darkness, the rite also symbolizes the supremacy of the lighter skinned Aryan race over the darker skinned people whom they encountered during their incursions into northwestern India. Included in the ritual is a comic passage of verbal abuse between a Brahman student and a courtesan, the aim of which may have been to assure fertility. Vedic rituals could have served as the foundation for the emergence of drama and theatre in India in much the same way that the Catholic mass is said to have paved the way for the reemergence of drama in medieval Europe. But it is not possible to ascribe a date to this transformation or to point to a specific event which may have caused it to occur.

Epic sources

The search for the existence of drama and theatre at some specific date in ancient history inevitably leads to the vast body of literature generated after the composition of the *Ṛg Veda,* in particular to the two great epics composed between about 1000 and 100 B.C. The *Mahābhārata,* the most voluminous of India's epics, often referred to as an encyclopedia of Indian thought, makes reference to performers *(naṭa),* among whom actors may well have been classed. In later history, the word *naṭa* was used to describe dancers and pantomimists, as well as actors. Unfortunately the epic does not describe the precise work of these ancient performers. The *Rāmāyaṇa,* India's other major epic, also makes reference to *naṭa,* and in addition, it speaks of *nartaka* who speak *nāṭakā.* The latter term is used later to refer to a specific type of drama. Yet there is nothing in the *Rāmāyaṇa* which proves the existence of Sanskrit theatre at this point in time.

Other Texts

In the oldest surviving text of Sanskrit grammar, composed by Pāṇini in approximately the sixth century B.C., reference is made to the *Naṭasūtra*, a textbook for *naṭa*, ascribed to Śilālin and Kṛṣāśva. Unfortunately the *Naṭasūtra* has not survived and Pāṇini does not describe its contents or give us any details about the authors of the work.

Greek Influence

Owing to the diligence of Greek historians, we know that in the early fourth century B.C. Alexander the Great marched into northwestern India, where he encountered and often did battle with numerous Indian kings and princes. It has been argued that troupes of actors normally traveled along to entertain him. Some scholars have argued that Indian monarchs may have seen Greek plays, been impressed, and commissioned their own bards to compose comparable works in Sanskrit. Scraps of evidence have been amassed in support of this so-called Greek influence. To support this argument the following suppositions have been made: The Sanskrit word for curtain *(yavanikā)* seems to have been derived from the word for foreigner *(yavana)*, a term the ancient Indians reputedly used to describe Ionian Greeks as well as numerous other foreigners. Keith quotes Windisch, the German Sanskrit scholar, who contends that a curtain took the place of the painted scenery at the back of the Greek stage. He also argues that Greek New Comedy, which flourished about the same time as Alexander's exploits, influenced the plot construction, development of character types, and themes of Indian plays (1924, 57-68).

In particular, *The Little Clay Cart (Mṛcchakaṭikā)*, one of the most famous Sanskrit plays, bears a striking resemblance to the Greek New Comedy. But no one has yet been able to prove that actors accompanied Alexander on his Indian campaign. We have no conclusive proof concerning the origin of the word *yavanikā*, nor can we say that *The Little Clay Cart* was derived from one or several Greek New Comedies or even when it was written!

Political Unity Minus Theatre

Drama and theatre are not mentioned during the first great Indian empire, that of King Aśoka (272-232 B.C.). In his public declarations Aśoka condemns the holding of public meetings, which are known from later evidence to have had performers *(naṭa)* as participants. Following this period in India's development, there was a sharp decline in empire and a diffusion of political unity throughout the country. It is during this period that definite signs of the existence of theatre emerge.

Concrete Evidence

To illustrate a grammatical point, Patañjali in his *Mahābhāṣya*, composed approximately 140 B.C., lists various methods by which events such as Kṛṣṇa's slaying of the wicked King Kaṁsa and the binding of Bāli may be described. Patañjali asserts that these dramatic actions could be depicted in one of several ways: (1) by pantomimic action, (2) through pictures which illustrate the story, or (3) through words alone. Although there is nothing in Patanjali's remarks to prove the existence of drama, all the elements are present for its formation: (1) acting in pantomime, if not with words, (2) recitation, (3) *naṭa* who could sing as well as recite, and (4) female dancers. Based on this evidence, Keith has made the point that like the Greek drama, which is said to have had its roots in competition between the forces of good and those of evil, the Indian drama may have arisen from the ritual competition between Kṛṣṇa, the personification of good, and Kaṁsa, the personification of evil. He links this with ancient vegetation rituals in which the out-worn spirit (symbolized by Kaṁsa) is destroyed by the new spirit (Kṛṣṇa) (1924, 36-46).

Miscellaneous Views

Diverse other arguments have been advanced concerning the origin of Sanskrit drama and theatre. Sir William Ridgeway theorizes that drama is the outcome of plays in which reverence is paid to the spirits of the dead, and that they were first composed to honor the spirits of the dead heroes who were personified by actors. Other scholars have contended that there were shadow plays and puppet shows in ancient India, and that they dictated the formation of drama. Some suggest that ancient pantomime coupled with epic recitation had a great deal to do with the way the theatre was formed. One proposal claims that the recitation of epic stanzas and the competitions that were held among skilled reciters led to the development of drama. The argument has also been advanced that theatre was first conceived in humble origins, and that plays were first composed in regional languages. When the Aryans came and superimposed their own language—Sanskrit—on the local inhabitants they adopted the regional-language plays and techniques and retained the regional languages (Prakrits) as secondary languages in the drama. Although interesting, the evidence supporting these conjectures is so slight that it is not possible to give them serious consideration here. Suffice it to say that the earliest plays which have been discovered (from about the first century A.D.) indicate that Sanskrit theatre was already fully formed and highly sophisticated. What intervening changes occurred between the first century B.C. and the first century A.D. which resulted in that sophistication are still a mystery.

Works Cited

Ghosh, Manmohan. Trans. and ed. 1967. *Nāṭyaśāstra.* Ascribed to Bharatamuni. 2nd rev. ed. Vol. 1. Calcutta: Manisha Granthalaya.

———. 1961. *Nāṭyaśāstra.* Ascribed to Bharatamuni. Vol. 2. Calcutta: Asiatic Society.

Griffith, Ralph T. H. Trans. 1963. *Hymns of the Ṛg Veda.* 4th ed. Vol. 2. Varanasi: Chowkhamba Sanskrit Series.

Keith, A. Berriedale. 1924. *The Sanskrit Drama.* London: Oxford University Press.

Additional Reading

Byrski, M. Christopher. 1974. *Concept of Ancient Indian Theatre.* Delhi: Munshiram Manoharlal.

Konow, Sten. 1969. *The Indian Drama.* Calcutta: General Printers and Publishers.

Shekhar, I. 1977. *Sanskrit Drama: Its Origin and Decline.* Delhi: Munshiram Manoharlal.

Characteristics of Sanskrit Theatre and Drama

Farley P. Richmond

What we know of the dramatic composition and staging techniques of ancient India is derived primarily from two sources—the extant plays with their stage directions and the numerous dramaturgical texts and books of theory and criticism. Unlike the Greek theatre, whose history relies at least in part on the ruins of theatre monuments, the Indian theatre does not have any surviving theatre buildings which might shed light on its origins and development. Also we are not privy to extensive production records which could help us to understand the performance events of ancient times. Yet a sufficiently large body of information remains from which some basic conclusions may be drawn.

From the plays we learn that Sanskrit theatre was sophisticated, if at times somewhat uneven in quality. Some of the finest poetry of the ancient world may be found in Sanskrit drama, and several of the extant plays compare favorably with dramatic literature elsewhere in the ancient world. A survey of the plays and dramaturgical texts leads us to believe that the theatre was conventional. That is, it called upon on actors and spectators alike to understand and decipher a complex code of gestures, movement patterns, and vocal expressions. Sanskrit theatre also was multidimensional. Dance and music contributed to the overall effect, although the extent of their interaction with dialogue and verse is far from certain. This was a theatre of imagination; costume and makeup were stylized and symbolic, not replicas of those of everyday life or of a particular historical period, and scenery was used sparingly, if ever, in the staging of plays. Special buildings appear to have been constructed for performances. Theatre had a sacred significance. Numerous rituals accompanied the construction of playhouses; the stage was consecrated before performance; and, from all the evidence, theatre served a religious function, as well as being entertaining and educational.

At no other time in India's long history has one genre of theatre been so universally understood and imitated throughout the subcontinent. This was primarily due to the predominance of the Sanskrit language, which was used at court and in religious rituals. Various forms of Prakrits (local languages and dialects) are also found in the surviving works. But Sanskrit was the language of the major male characters; and although the theatre of this period thus had a multilingual character, it is regarded as Sanskrit, rather than Prakrit, theatre. Sanskrit theatre was the national theatre of its day, reflecting the dominance throughout India of the civilization represented in the plays. As a court- and temple-supported art, theatre was exclusive rather than popular in its appeal.

Although Asia boasts many other ancient civilizations, India seems to have been the first country of the region to support an active, sophisticated theatre. And yet theatre in India came into being at a relatively late date, compared to the development of theatre in Greece. However, once theatre finally took shape in India about the second century B.C., it continued virtually uninterrupted for at least twelve hundred years, and when it came to an end it did not disappear overnight. Beginning around the tenth century A.D. no more significant dramatic poets appeared on the scene, but from time to time thereafter, authors tried their hand at writing Sanskrit plays in the style of the acknowledged classics. Even today one may still find dramatic works composed in Sanskrit, but they are either derivative or are modeled on the conventions of Western playwriting.

In the following pages I will explore the characteristics of theatre and drama in ancient India, including the *Nāṭyaśāstra,* the principal text of ancient dramaturgy; the theatre organization; acting; music and dance; the spectators and occasion of performance; the physical theatre, stage, and function; playwrights and their works; and the aesthetics of dramatic composition before concluding the chapter with a brief examination of some of the possible reasons for the decline of the ancient theatre.

Nāṭyaśāstra

The most important text of dramaturgy that ancient India produced is the *Nāṭyaśāstra.* The title literally means "drama science" but a cursory review of the text is all one needs to indicate that this is more than a science of drama; it is a veritable encyclopedia of knowledge concerning Sanskrit drama and theatre, the most comprehensive work of its kind to survive anywhere from ancient times. Without the benefit of the information it contains, we would know relatively little about a number of important issues concerning the Sanskrit theatre. And yet the work is an enigma, prompting a host of questions which have not been and perhaps may never be answered.

Scholars differ on the date of the composition of the *Nāṭyaśāstra*. Based on internal evidence, it is generally believed that it was composed during the period between 200 B.C. and 200 A.D. Although the authorship is traditionally attributed to the sage Bharata, an unevenness of writing styles has led some critics to conclude that it is the work of several hands.

At a glance the *Nāṭyaśāstra* is obviously broader in scope than the other great text of ancient dramaturgy, Aristotle's *Poetics*. The work consists of numerous chapters, beginning with the origins of theatre. Within its scope are a multitude of topics, such as theatre architecture, acting, costuming, makeup, properties, dance, music, play construction, poetic composition, grammar, composition of theatre companies, audiences, dramatic competitions, the actor community, and ritual observances, to name the more important.

The text has exerted a profound influence over the formation and structure of successive theatre genres throughout India. Our understanding of the musical system of ancient India is based almost entirely on the *Nāṭyaśāstra*. Most forms of Indian dance are said to derive their language of gestures, steps, poses, and movement patterns from the *Nāṭyaśāstra*. By comparing the rules of play construction outlined therein with the works of several generations of playwrights, scholars have built a considerable body of dramatic criticism. In attempting to interpret the various aspects of the *Nāṭyaśāstra*, a whole body of theoretical texts has come down to us from ancient times. And debate over the meaning of certain important words and phrases still occupies the attention of scholars throughout the world.

Despite the fact that the *Nāṭyaśāstra* has been the center of scholarly debate and the source of practical information for centuries, the extant manuscripts that are known to exist in private collections and libraries throughout the world have never been compiled into a definitive edition.

Theatre Organization

Relatively little is known about theatre companies in ancient India, although there are tantalizing scraps of evidence in the *Nāṭyaśāstra* and various other dramaturgical sources, historical reference works, and plays to suggest a possible hierarchy, composition, and manner in which groups practiced their occupation. The first company seems to have been a family affair. The leadership was in the hands of a family elder (Bharata) who received his knowledge and training directly from Brahmā, the creator god, and who in turn passed on his wisdom to his sons (one hundred in all!). As his contribution to posterity, Bharata gave the world the *Nāṭyaśāstra,* which is reputed to be a record of his teachings as he learned them from Brahmā. Whether theatre continued to be a family affair throughout its long history is unclear. It does appear certain that companies had a group leader, actors of various ranks of importance who received systematic training, numerous assistants and minor functionaries, and perhaps separate categories of musicians and dancers and even playwrights.

After briefly reviewing the activities, means of support, and social status of the group leaders, actors, minor functionaries, and musicians and dancers I shall address the characteristics of ancient acting and what little is known of music and dance of the period.

Leadership

At the head of the company was the troupe leader, who also may have served as producer as well as stage manager *(sūtradhāra).* His major job seems to have been supervisory, although there are references that suggest that he was also an actor, perhaps even the principal actor, of the company and took the roles of the leading male characters.

As leader of the company his responsibilities may have included establishing engagements for the performance of plays with officials of a temple, the court, or wealthy citizens. As stage manager—the original Sanskrit term literally means "holder of the thread or strings"—he may well have been the architect of the ancient theatre buildings. As the producer of the play, the *Nāṭyaśāstra* charges him to "pay attention to the feelings *[bhāva]* gestures *[mudrās]*, and the *sāttvas* [perhaps 'feelings'] in representing the psychological states *[bhāva]* through various characters (that may appear in drama)" (Ghosh 1967, 513).[1]

Other evidence suggests that the group leader served a similar function to that of the modern stage director. The *Nāṭyaśāstra* assigns him the job of distributing roles in a play and of teaching the rules which apply to the art and craft of the actors and actresses. It goes further and lays down the characteristics of an ideal leader. He should be a person capable of fulfilling all these diverse and difficult responsibilities, who has an excellent memory and intelligence, and who is versed in music, language, astrology, and physiognomy, as well as theatre. He should be sensitive to human frailty and familiar with the customs of different regions of the subcontinent.

The troupe leader seems to have presided over a company of actors, singers, dancers, musicians, backstage workers, and playwrights. We know a great deal more about the actors and playwrights than we do about any of the other functionaries.

Actors

The generic term for actor was *bharata,* which seems to be the hereditary group to which they belonged. Actors were also called *naṭa* or *nartaka* and actresses were known *as naṭī* or *naṭakaya.* In the literature of the time, these terms were also used interchangeably to describe dancers and acrobats.

Apparently both men and women could perform in the plays. Roles could be taken by actors who were the approximate age and sex of the characters. However, younger actors could also play older characters and older actors could perform younger roles, and men were permitted to play women and women could act the role of men. There are even records which indicate that entire companies were composed of women. Actresses were usually regarded as better suited than men to sing, but were not thought fit to participate in battle scenes, as actors were regarded as better suited to depict the sentiments of heroism and rage. Men's voices were thought best for recitation, because they were regarded as stronger than those of women.

If the *Nāṭyaśāstra* is correct in its assessment, the knowledge required of an actor must have been considerable. The ideal actor was assumed to have intelligence, strength, physical beauty, knowledge of time and tempo, and appreciation of sentiments and emotions. He must be the proper age to play a role, possess curiosity, be able to acquire knowledge and other arts, and have a retentive memory. He must understand the vocal music that prompts dance, suppress stage fright, and display enthusiasm. He must sing, dance, use complicated, symbolic hand gestures, and be able to recite the text clearly enough to be heard throughout the theatre. Since critics were supposed to judge the merits and faults of his work, he was obviously obliged to pay close attention to the accuracy of his execution.

According to the *Nāṭyaśāstra,* actors could be faulted for the following: (1) unnaturalness in acting, (2) incorrect movement, (3) unsuitability for a role, (4) forgetfulness, (5) improper use of gestures, (6) defect in costume and ornaments, (7) detect in rhythm of execution, (8) improper projection, and (9) excessive laughter or weeping. Like a yogi who is said to possess the power to transfer his soul from his body into that of a person who has recently died, the actor was supposed to think of himself, "I am he" (Ghosh 1961, 216-217).

Procedures for training actors are also established in the mythological origin of the theatre. Instruction came directly to people from god via a wise sage. Thus, a drama teacher is an older person, usually a father, who instructs younger persons, usually his sons. The *Nāṭyaśāstra* seems to have been designed to serve as a guide to the correct procedures. In this way ancient wisdoin was preserved and passed down from generation to gentration. When the actor had no excess to a text or a competent teacher, the text recommends that he observe and imitate the practices of others. Today, many of the same training procedures proposed by the *Nāṭyaśāstra* are practiced by performers of various rural art forms and even by the actors of some urban theatre schools and companies.

The *Nāṭyaśāstra* charges actors to keep physically fit by undergoing a regular routine of exercises. The exercises were to begin with a purgation of the bowels followed by special dietary restrictions during the period of exercise. The body was to be massaged with sesame oil or with barley gruel after exercising. The proper place for exercises was the floor, which was called the "mother" of exercise. The regimen included leaping in the air and stretching, accompanied by special movement patterns, perhaps drawn from those used in the preliminary rituals before the performance of a play. Training must have taken many long years of hard work, starting with the apprenticeship at a very young age of the pupil to a teacher.

Actors may well have specialized in playing specific roles according to set role designations. The hero, heroine, and clown are a few of the more important categories of roles. It is possible that actors who had a natural aptitude, as well as skill and training, gravitated to one of these categories. Such a system encouraged specialization, with the more difficult parts going to the better actors of the troupe and the smaller roles assigned to younger, less experienced players. Given this hierarchy, it is entirely possible that the actors who were senior and more experienced enjoyed a greater share of the profits. Since the society venerated its elders, the practice probably did not seem unfair to highly talented but less experienced younger players.

Other Company Members

The stage manager is thought to have had an assistant whose job was to help in the prologue and recitation of the benediction. He may also have aided in the preparation and construction of the theatre building. References in the prologues of many plays in which actresses appear lead us to believe that the actress and the stage manager were husband and wife, although this has not yet been established as anything more than a theatrical convention. Numerous other assistants seem to have shared in the backstage responsibilities, including makeup artists, dressers, makers of ornaments and garlands, painters and dyers and modelers of effigies, and washermen.

A company may have had musicians at its disposal, but it is also possible that the musicians constituted a separate group of artists who performed with different troupes of players depending on the occasion and the agreements that had been struck. Musical ensembles are thought to have varied in number from small to medium or large, depending on the particular needs of a performance. The ensembles consisted of male and female vocalists, flutists, players of string instruments, drummers, and those who kept time with cymbals and bells.

Theatre companies may also well have included a separate group of dancers, engaged to perform in the complicated preliminary rituals and perhaps between acts or at places undesignated in the extant playscripts. However, it is also possible that the actors were responsible for performing their own dances in the show.

Means of Support and Social Status

Apparently the performers were sustained by payments made by their patrons. This might have been in the form of simple goods, like rice and oil to be used in preparation of the meals, as well as gifts of clothing and precious metals or jewels. When particularly moved by a performance, spectators might show their appreciation by offering material rewards to the players. It does not seem to have been the practice to sell tickets to a play or to take up a collection from the spectators.

The social position of the players does not appear to have been very high. Numerous ancient writers speak of actors in the same breath as prostitutes and bandits. To prevent themselves from being ritually polluted, Brahmans were never to take food from the hands of an actor. It was a common belief that the actors sold their wives as prostitutes, yet some historical authorities indicate that close personal ties existed between some actors and their royal patrons. Most probably acting companies varied widely in terms of their standing in the community and among the various social classes; some probably enjoyed a higher degree of respectability than others.

In the last chapter of the *Nāṭyaśāstra,* Bharata describes the descent of his line from that of high caste Brahmans to lowly Sudras. According to the account, Bharata's sons caricatured the sages in a particular type of drama which caused others to laugh. In retaliation the sages cursed the boys and' those born of their line to be Sudras, the lowest rank of the caste system. Fearing that this might mean the end of the dramatic art, India and the other gods approached Brhmā. The sages reassured the gods that drama would continue even if the actors were cursed. However, Bharata's sons were so miserable that they threatened to commit suicide. Just then, so the story goes, a king requested that Bharata and his sons descend to earth and perform plays at his court. Thus drama was saved, and consequently it fell within the province of kings, who took it upon themselves to patronize theatre and drama as one of their royal obligations. Bharata's sons married earthly women and drama was sustained and the art passed down from generation to generation.

Acting

The acting techniques that were practiced in ancient India during a major period of its theatre history are lost to us, except for what we may glean from written materials. And yet there is cause for some rejoicing. Compared with the evidence concerning Western acting practices, the body of information in India is vast, though at times it may seem cold, since acting is a temporal art and it is difficult, if not impossible, to place performances within the framework of a civilization thousands of years old and attempt to understand it within that perspective, primarily on the basis of surviving texts. Most of what we know of the Indian system of acting may be found in the pages of the *Nāṭyaśāstra.*

It may be that there were two main schools of acting in ancient India. The *Nāṭyaśāstra* recognizes two styles of representation *(dharmi)* in dramatic performance: the realistic *(lokadharmī,* literally meaning "the popular"), and the conventional *(nāṭyadharmī,* "the theatrical"). To what extent realism as we know it today actually existed on the stage is only a matter of conjecture, for the *Nāṭyaśāstra* almost exclusively emphasizes conventional techniques.

In addition, four styles of performance are identified: the verbal, which probably placed emphasis on speech alone; the grand, which may have emphasized spectacle; the graceful, perhaps referring to an emphasis on feminine movements and music; and the energetic, which may have referred to masculine movement and forceful music.

A tantalizing reference in the *Nāṭyaśāstra* also implies that dramatic performances may have varied from one geographical locale to another. No matter where one went in ancient India, whether north, south, east, or west, there are said to have been local characteristics which prevailed.

The term acting *(abhinaya)* literally means "carrying toward," implying that the actor is one who brings a performance to the spectators. According to the *Nāṭyaśāstra,* the art of acting consists of four major elements: bodily movement, voice, spectacle, and sentiment or emotion. I shall briefly summarize these aspects as they are presented in the text.

Bodily Movement

Five long chapters of the *Nāṭyaśāstra* are devoted to a comprehensive survey of bodily movement, more than is given to any other aspect of acting. For this reason, bodily movement may well have been considered the most important component of the actor's art. Unfortunately the chapters which deal with the subject provide no visual illustrations. Sometimes this leads to a monumental problem of interpretation.

Which one of the several modes of dramatic entertainment the author is referring to is also unclear. At least three models are recognized—pure dance *(nṛtta)*, pantomimic dance *(nṛtya)*, and drama *(nāṭya)*. As indicated in the general introduction to this book, in parlance, pure dance refers to movements, gestures, and facial expressions which are abstract, which do not have a representational context, and in which there is an instrumental accompaniment or abstract syllables recited by a singer. This mode of physical expression glorifies the skill and technique of those who dance in strict patterns, according to the dictates of time and space. Pantomimic dance, as the name implies, stresses the enactment of a story. In many forms of Indian dance today the dancer usually performs a short scene or episode in which gestures and facial expressions rather than dance steps or movements are the important features. A typical example is the dressing and makeup preparations of a beautiful woman waiting for her beloved. One of the musicians usually sings a text to accompany the dancer, who sometimes mouths the words. We are not sure to what extent pure dance and pantomimic dance were combined in the drama. The *Nāṭyaśāstra* seems to recognize several kinds of dramatic expression under the broad category of drama: dance drama, literary drama, and musical drama. Unfortunately it does not distinguish clearly how these differ from one another and how movement and gestures are to be used in each.

Bodily movements are said to be of three kinds: those of the limbs, those of the face, and those of the entire body. The *Nāṭyaśāstra* identifies the major limbs as the head, hands, breasts, sides, waist, and feet. The minor limbs are those of the face—the eyes, eyebrows, nose, lower lip, and chin. Movements of the minor limbs are discussed first in the text and may well have been the more important in communicating the sentiment of a play. Thirteen head positions are itemized and thirty-six glances, each categorized according to the major sentiments and emotions. Movements of the eyes, eyelids, and eyebrows are categorized and identified according to the emotions they convey. The nose, cheeks, lower lip, chin, mouth, and neck are all discussed according to their ability to communicate emotional meaning. In addition to makeup, the *Nāṭyaśāstra* identifies facial color as an important aspect of expressing emotion.

An entire chapter of the text identifies gestures appropriate for one as well as for both hands. It also cites gestures appropriate for use in pure dance. Various meanings are assigned to each gesture. If the gesture language is as complicated as it is in surviving dance and dance-drama forms, then it would require the help of an expert teacher to learn the precise way the gestures were to be executed.

In another chapter precise bodily movements are prescribed to correspond to emotional states. Positions of the breast, side, stomach, waist, thigh, shank, and feet are categorized and discussed at length.

The specific poses and physical reactions of actors in various situations are referred to as *cārī*. Thirty-two *cārī* are identified; approximately half of them are referred to as earthly and the remaining are called aerial. Precise stances are also identified and designated as appropriate for specific situations. The wielding of weapons, such as the shield and bow, are encompassed within the discussion of the poses. In a separate chapter, the *Nāṭyaśāstra* describes the movements which are to accompany the poses. Most of these movements refer to fighting in personal combat.

Chapter thirteen discusses at some length the gaits which are appropriate for various characters, depending on their sex, rank, age, and temperament. It provides a particularly vivid example of the gait appropriate for a blind man or one walking as though in darkness. It says, "The actor's feet glide over the ground and hands grope for the way" (Ghosh 1967, 225). From this description it would appear that this movement was firmly rooted in reality. The extent to which physical movements were conventionalized cannot be determined from the text.

Different eye, head, face, and hand movements are designated for each of the five senses. Provisions are also made for the actors to show three broad categories of reactions to the object of a situation—

favorable, unfavorable, or indifferent. Other conventional patterns of movement are discussed with reference to the walking around *(parikramaṇa),* traversing from one locale to another. The stage directions of extant texts provide numerous examples that suggest that acting was highly conventional and yet had a firm basis in reality.

Voice

Five chapters of the *Nāṭyaśāstra* are concerned with language, but relatively little space is devoted to the use an actor should make of his voice in performing a play. Most of what is said applies to the literary composition of a dramatic work. Chapter fifteen begins with a discussion of the human voice, but soon digresses to a consideration of questions of Sanskrit grammar and language. Chapter sixteen identifies the various metrical patterns that should be employed in composing verses, and seventeen deals with the appropriate composition of speech in various dramatic situations, and eventually becomes a linguistic lecture with examples.

The beginning of chapter eighteen concerns Prakrit grammar and language. Some of the more important points about speech in the plays are made shortly thereafter. Plays are said to be composed of two languages: refined *(saṃskṛta)* and vulgar *(prākṛta).* Sanskrit is regarded as appropriate for the dialogue and verses of gods and kings. There is also a common language and even a language for animals and birds. Rules are set forth regarding the kinds of characters that may use various languages to converse with each other. Regional dialects are cited as appropriate for various characters according to their rank and station in life. The point is also made that "the producer of the plays may however at their option use local dialects for plays may be written in different regions [for local productions]" (Ghosh 1967, 329). Chapter nineteen concentrates on the appropriateness of various modes of address as determined by the character's occupation and whom he is addressing. The chapter ends with a brief description of vocal registers, accents, intonations, vocal qualities, enunciation, and the uses of the pause.

Spectacle

The element of spectacle is divided into four component parts: stage properties, decorations, makeup, and "creatures." With reference to properties, realistic objects such as swords, spears, bows and arrows, and the like are considered inappropriate for stage use. Properties resembling these objects were to be fashioned of lightweight material and painted to look like the actual object. Great care was required in the construction of Indra's banner pole, which was used as part of the preliminary rituals.

Among accessories, the *Nāṭyaśāstra* classifies mountains, temples, caves, idols of deities, elephants, horses, aerial vehicles, and houses. It is possible that these objects were used to symbolize an actual place and thereby establish the identity of a certain part of the stage. The items were to be "fashioned in cane frame" and "covered with cloth." This suggests two possible methods of construction. One possibility was that the object was constructed of bamboo and wrapped with cloth, producing an outline of the object. Properties constructed in this manner are found in the Noh theatre of Japan. The second possibility is that cloth was stretched over a bamboo frame and then painted to resemble the object. This method of construction is used today in the *aṅkīya nāṭ* of Assam. The text also makes reference to the construction of masks.

Elsewhere in the *Nāṭyaśāstra* we find mention of a curtain *(yavanikā)* which may have been held by attendants and then lowered or pulled aside to reveal a character. Some scholars have suggested that curtained entrances from the dressing room or traverse curtains stretched between the stage pillars were regularly in use. The practice of employing a curtained entrance, as in the Noh of Japan and *jingxi* of China, may have been common in the Sanskrit theatre as well.

The *Nāṭyaśāstra* specifies garlands, ornaments, and costumes as items of decoration for an actor. Specific instructions are given for the decoration of men and women according to the region of the country to which they belonged, their caste, station in life, and the occasion of their appearance on the stage. Headgear, earrings, necklaces, bangles for the arms, wrists, and ankles, and finger and toe rings are also mentioned. These were grouped according to whether they were to pierce the limb, be tied up, hang, or circle the body. From the tops of their heads to the tips of their toes the actors of the Sanskrit theatre were to be profusely and brilliantly ornamented and costumed. Heavy objects such as gold or precious stones were deemed inappropriate because they might easily tire the actor due to their excessive weight. Instead the *Nāṭyaśāstra* advises that they be constructed of light wood and glazed with shellac to resemble the real object.

No clear indication is made in the text concerning the reproduction of historically accurate costumes, ornaments, and properties. Some scholars believe that the actors wore the garb of the particular period in which the play was written. Others think they dressed according to the dictates of the period in which the play was set. Still others regard the decorations as fanciful, conforming to no particular period or style.

The *Nāṭyaśāstra* suggests that makeup be applied to the face and body of the actor according to the region of the country, caste, age, and occupation of the character. Specific colors are designated for specific character types, though it is not clear if the colors prescribed in the text were conventional or realistic.

The meaning of "creatures" (*saṃjīva*) is not made clear in the *Nāṭyaśāstra*. It may be that it refers to actors who portray animals, birds, and super-human beings, or it may refer to actors who manipulate giant effigies like those used in the *aṇkīya nāṭ* of Assam.

Sentiment and Emotion

Critics have not yet come to an agreement concerning the precise definition of two important aesthetic terms: *sāmānābhinaya*, which seems to be a synonym for *sāttvikābhinaya*. In chapter twenty-four of the *Nāṭyaśāstra, sāttva* is defined as "something invisible" and is said to be the basis of the dramatic production (Ghosh 1967, 442). The whole of the chapter is concerned with the decorous behavior of the characters, particularly the female characters of the plays. Special attention is lavished on the correct speech or gestures appropriate for various occasions. This has led at least one critic to conclude that the term means an "expression in a graceful or charming manner of the various *bhāvas,* i.e. feelings (of a character)" (Rangacharya 1971, 36). Another critic thinks *sāttvika* refers to the psyche of the character. Still another scholar believes that the author of the *Nāṭyaśāstra* meant to introduce a realistic style of acting which depends on correct social behavior as its guide, in contrast to the conventional school of acting mentioned in the majority of the chapters. Whatever the correct definition of the term, the *Nāṭyaśāstra* has presented a notion which requires of a performer more than the external representation of character, such as correct movements, speech, and ornamentation. It implies that the standard of stage performance goes far beyond the mastery of technical considerations, and that the players be believable in order to satisfy the spectators.

Music and Dance

Examining the texts of the extant Sanskrit plays, it is next to impossible to determine how music and dance were integrated into a performance. Indeed from the plays alone, you might well get the impression that Sanskrit drama was dialogue drama without the benefit of music and dance, except for the rare occasion in certain texts in which a character is said to sing or dance or do both. And yet the *Nāṭyaśāstra* tells us something quite different. Songs and dances seem to have been very much a part of the total performance experience.

As indicated earlier, the chapters devoted to music are our earliest source of information about the music of ancient India. Some scholars contend that the music which we now conventionally associate with India is relatively recent in origin and that the *Nāṭyaśāstra* refers to a musical system which was unique and which has undergone considerable changes with the passing of time.

According to the text, melodic sounds were either produced by the human voice or by musical instruments (literally by the *vīṇa*, an ancient stringed instrument). Instruments traditionally associated with the drama are either stringed; covered, namely the various drums; solid, such as the cymbal; or hollow, like the flute. Some scholars contend that the flute was the leading musical instrument, for it provided the pitch from which the singers took their cues and served a similar function to the present-day drone instruments, such as the tambura and harmonium.

The appropriate place for the musicians to sit was upstage center, between the two doors leading to the dressing room. They may have sat on a rug on the floor and watched the progress of the play from this strategic vantage point. This gave them a full view of the acting area and also provided easy access to the dressing room in case changes in format needed to be made on short notice.

It is apparent from the early chapters of the *Nāṭyaśāstra* that music was essential to the progress of the preliminary rituals and to the consecration of the theatre building. The songs that were sometimes sung by an actress in the prologue probably needed a musical accompaniment, as did the songs referred to on occasion in some of the scripts. A few scholars contend that actors sang some, if not all, of the verses of a play. If this were so, then these too would have required musical accompaniment. According to other scholars, the actors recited or chanted the verses. This controversy has not yet been resolved. Many of the plays indicate that classical and folk dances were to be performed as part of the action. They too would have required a musical accompaniment.

We learn from the *Nāṭyaśāstra* that songs were used extensively throughout a play, for various purposes. The songs were composed by the musicians, and their content was guided by the content of the verses of the play. Songs were used: (1) to introduce a character when he or she first comes on the stage, (2) when a character exits during the middle or at the end of an act, (3) to reinforce a mood already established in the dramatic action, (4) to change the mood of a scene, or when a situation changes, and finally, (5) when some gap occurs in the action due to a mishap, such as the need to adjust or replace a costume or when an overpowering feeling causes great tension.

Apparently, the songs were composed in Prakrit and had a symbolic purpose. It is thought that they were sometimes sung behind a curtain before an entrance, perhaps even by the actor who was preparing to enter. It is also assumed that the melody was rendered completely through at least once, and then was repeated to the accompaniment of the rhythmic pattern of the drums. Since none of the extant plays include references to the songs, nor do they indicate how they were to be incorporated, it is a matter of pure speculation what their effect might have been on the total production.

Although dance is mentioned in a few plays, the details of its execution are also a matter of conjecture. It would seem that its main function was to serve as a part of the preliminary rituals and to heighten the dramatic impact of a particular scene. It is possible that dancelike movements and gestures were used by the actors throughout a performance as part of their acting technique. Practices such as this are still to be seen in the *kūṭiyaṭṭam* of Kerala, which I will discuss in the next chapter. Perhaps these practices provide a clue as to the authentic style of performance practiced at least in the declining years of the Sanskrit theatre.

Spectators and the Occasion of Performance

The *Nāṭyaśāstra* speaks of those who attend the theatre as spectators (*prekṣaka*, literally "those who see"). From this it would seem that the ancient Indian theatre laid particular stress on the visual aspects of performance, and yet there is ample evidence that both sight and sound were to be an effective blend and that the spectators were to be equally cognizant of the importance of both.

Actually very little is known about the individuals and groups that attended plays in ancient India. In the prologues to many of the plays, a stage manager speaks directly to the spectators and flatters them by declaring that they are learned and wise. Perhaps what the stage manager says was true and not just a theatrical convention; however, it is more than likely that those who attended Sanskrit plays varied widely in learning, judgment, and taste. It may be remembered that in the first chapter of the *Nāṭyaśāstra* Indra asked Brahmā to create drama for the amusement and edification of all the color or caste groups *(varṇa)*, from the highest caste Brahmans, arbiters of ancient knowledge and wisdom, to the lowly Sudras, simple peasants, as well as for all the occupational groups in between.

Conforming to the general practice of specifying ideal types and classes of individuals and things, the *Nāṭyaśāstra* identifies the characteristics that the ideal spectators should possess. They should be of good character, high birth, quiet and learned, partial, advanced in age, alert, honest and virtuous. The audience was to be proficient in drama and acting, expert in playing musical instruments, knowledgeable about costumes and makeup, and expert in other branches of the arts and crafts. They should understand the meaning of the sciences *(śāstra)*. Grammar, prosody, sentiment and emotion—all these were to be known and appreciated by the ideal spectator.

Occasion of Performance

The occasion for a theatre performance arose from a desire on the part of an individual, group, or institution to organize a public gathering to celebrate some great occasion; perhaps the first and most important occasion was the celebration of a temple festival. Since theatre architecture was considered a branch of temple architecture, it would appear reasonable that performances in the temple theatre were given on a regular basis when astrological conjunctions favored honoring some deity or other. Indeed the temple deity may well have been the chief guest at such occasions, if not in the form of the temple idol, at least in spirit. After all the *Nāṭyaśāstra* describes the performance of a play as a sacrifice to the gods. The numerous rituals accompanying the building of the theatre, the stages in its construction, its consecration, and the preliminaries preceding the performance of a play strengthen the contention that dramatic events were thought of as sacred affairs.

Other evidence suggests that plays were performed for the coronation of kings, marriages, the birth of a son, the return of a traveler, and the celebration of the confiscation of a town or a state. The *Nāṭyaśāstra* asserts that "of all the duties of the king, this [the performance of a play] has been proclaimed as possessing the best result. Of all kinds of charities, allowing people to enjoy a dramatic show without payment, has been praised most" (Ghosh 1961, 237).

Positive and Negative Critical Responses

The *Nāṭyaśāstra* also recognizes various conventional ways spectators were expected to respond to a production. When the comic sentiment dominated, a spectator was to show a slight smile, a half smile, or perhaps laugh uproariously. If well-acted, moments depicting virtue were to be accorded cries of "excellent." The sentiment of wonder was expected to prompt the spectators to sigh, "How wonderful!" And pathos could cause them to weep and mutter, "How pathetic!" Tumultuous applause greeted astonishing sights and the hair on their arms was to stand on end during moments of fear. In general the spectators were expected to become totally engrossed in the theatrical event and to respond to the acting accordingly. The extent of their involvement seems to have been conventionalized. For example, the stage manager in the prologue to Kālidāsa's *Śākuntala (Abhijñana-śākuntala)* remarks to the actress after she has completed her song, "Charmingly sung, good lady! Look, the entire audience seems like a picture, their souls entranced by your melody" (Kālidāsa 1964, 201).

The *Nāṭyaśāstra* says that a performance was judged to be a success in one of two ways—either it was considered a "human" success, prompting the reactions mentioned above, or it was considered a "divine" success. "Divine" successes seem to have been accorded to performances in which there was no noise, disturbance, or any unusual occurrences to distract from the discriminating depiction of the actor's emotions. Perhaps the *Nāṭyaśāstra* means that divine performances were those in which the spectators showed neither interest nor disinterest, in which they were elevated to a state of supreme bliss akin to that of a yogi who experiences divine revelation.

The *Nāṭyaśāstra* also enumerates the possible disasters that might befall a production. Natural disorders such as wind, fire, rain, lightning, and earthquakes, among others, are categorized as disasters caused by the gods. Man-made calamities which disrupt performances fall into two categories: (1) those brought about by jealous enemies who scream, throw things, and clap loudly; and (2) those caused by the actors, the details of which have already been discussed.

The *Nāṭyaśāstra* takes the sensible view that "nothing can be devoid of merit or free of fault" (Ghosh 1967, 522). And so there are those among the spectators who are recognized as critics and are assigned a place in front of the stage, to evaluate the proceedings and to award prizes. People skilled in any of ten types of activities are fit to be critics. They are: (1) the Brahmans, those expert in performance of sacrifice; (2) experts in acting; (3) prosodists; (4) grammarians; (5) kings; (6) archers; (7) painters; (8) courtesans; (9) musicians; and (10) courtiers. Each expert was assigned to judge the merits and faults of the performance according to his or her own level and area of expertise For example, the king was to judge the actors who portrayed kings, and so forth. Recorders had the task of keeping score of the faults noted by the critics during a show and tallying the results. Prizes were awarded to the actor who had accumulated the fewest faults. A banner was the usual prize. In case of a tie, the king was to choose between the two contestants, or he was given the privilege of awarding a banner to each, if he happened to be fond of both.

Theatre, Stage, and Function

Unfortunately, none of the theatre structures of ancient India have survived the ravages of time. To make matters worse, we have no visual evidence either—no drawings or floor plans, no sketches made by a diligent visitor, not even the foundations of a building in ruins. Our only substantial source of information regarding the size and shape of the buildings that are said to have housed Sanskrit play productions is the *Nāṭyaśāstra,* and from the discussion it appears to have been written with the intention of laying down rules for the construction of "ideal" models, rather than recording the size and shape of existing structures.

The *Nāṭyaśāstra* attributes Viśvakarmā, the heavenly architect, with conceiving of the first playhouse. As we learned earlier, he did so to accommodate a request made by Brahmā, who wanted a sanctified place where actors might perform a play uninterrupted by malevolent spirits. The text gives no details about the appearance of the "first" theatre except that construction was begun on an auspicious day and that the gods agreed to assist in its protection.

The text prescribes three shapes suitable for playhouses: rectangular, square, and triangular. Each was to come in a small, medium, and large size. Bharata lavishes more of his attention on the medium-sized rectangular building than he does on any other structure. He says that this is the ideal structure because spectators may easily see the facial expressions of the actors and hear their speeches and songs.

Characteristics of the Major Structure

The parameter of the medium-sized rectangular buildings was ninety-six feet by forty-eight feet; and according to directions, it was to be subdivided in half, thus creating two equal squares forty-eight by forty-eight feet. The space of one of the squares was allocated for use by the spectators. The other half was again subdivided and the area closest to the spectators was designated as the acting area and that at the extremity of the building was reserved for the dressing room.

Considering the limited size of the space reserved for the spectators, it would seem that relatively few people could have witnessed a performance of a Sanskrit play at any one time. Perhaps no more than five hundred persons, and possibly only two hundred patrons, comfortably seated, saw plays in the medium-sized rectangular playhouse. This obviously promoted a sense of intimacy between actors and audience inconceivable in the theatre structures of ancient Western cultures, such as those of Greece and Rome or even that of Elizabethan England.

The *Nāṭyaśāstra* specifies that special procedures be followed when selecting the site and clearing the land for construction of a new playhouse. Care was to be taken in laying out the basic divisions of the structure, otherwise disasters might befall the country and those responsible for causing the mistakes. With the sounding of conch shells and the playing of rhythmic instruments the foundation was laid. Undesirables were denied access to this ceremony, probably for fear of casting an evil eye on the proceedings. After the site was marked out on the ground, the gods of the ten directions were honored with ritual ceremonies.

The next step was the raising of the walls and pillars. Four pillars were designated to represent the four castes or color groups. This would seem to support the implication made in the first chapter that drama was created as a form of amusement and edification for all four castes. Symbolically the four castes were the pillars of the theatre.

Considerable disagreement exists among scholars over the meaning of the physical feature of the theatre building that was to be constructed next. Apparently, it had pillars and was higher than the level of the stage. Little more may be said for certain about its shape and function.

The notion that the playhouse was to resemble a mountain cave has led some critics to wonder if the early theatres were actually constructed in caves. Some years ago a cave was found in central India that the Archaeological Survey of India dubbed a theatre, but there is no proof that theatres were ever carved into the sides of mountains. It would seem that if this were a common practice in ancient times, other examples would still remain among the numerous cave sites of the subcontinent. It is more likely that the *Nāṭyaśāstra* meant to imply that the ceiling of the edifice was constructed in such a fashion that it acoustically aided the voices of the actors and the music and drums.

Small windows might have been inserted high up in the structure as ventilation and perhaps even to let in some light. The final step in the construction of the structure seems to have been the plastering, white-washing, and painting of the walls. The paintings were to depict scenes of "creepers, men, women and their amorous exploits" (Ghosh 1956, 30).

We know almost as little about the stage of the ancient Indian theatre as we do about that of Shakespeare's Globe or the Theatre Dionysus of fifth century B.C. Athens. Several points are fairly clear, however. The maximum potential acting area of the ideal structure was twenty-four by forty-eight feet. Two Sanskrit terms used to describe the stage have led some scholars to conclude that the stage was further subdivided in half, the part closest to the audience referred to as the "head of the stage" and the part nearer to the dressing room as the "seat of the stage." The front half is thought to have been slightly lower than the back half. A wall with two doors seems to have separated the dressing room from the acting area. As noted earlier, the space between the doors was reserved for the musicians. The stage seems to have been on a different level from that on which the spectators sat, thus it was clearly demarcated as an acting area. The floor was to be elevated and level as the surface of a mirror, suggesting that it may have been reflective. The *Nāṭyaśāstra* indicates that it was not to be convex or rough, like the back of a tortoise or a fish. Precious stones are said to have been laid in the foundation of the stage, probably for ritual rather than practical purposes. The stage had pillars, which either helped to support the superstructure of the roof of the main building or to support a separate roof which may have been constructed above the acting area. The text is uncertain about the exact number, placement, and function of the pillars. Wooden figures of elephants, tigers, and snakes were to be carved, possibly on the pillars or perhaps on the ceiling above the stage, if the stage had a separate ceiling.

Other Structures

The procedures for the construction of the square theatre are almost identical to those outlined for the construction of the middle-sized rectangular structure. A square was subdivided in half and the area relegated to the actors again subdivided in half, forming a dressing space and stage. Bharata describes a few unique details of this structure worth repeating here, since they may shed light on the features of the ideal structure. He says that the outer walls of the square theatre were to be constructed of bricks. Rows of seats for the spectators were to be made of wood and brick. The seats were to resemble a staircase, in that each successive row was higher than that which preceded it, potentially providing all the spectators with a clear view of the stage. Whether or not the seats were arranged in this manner in the medium-sized rectangular houses is still speculative.

Little of consequence may be said of the triangular theatre. Such a shape is uncommon in the history of architectural units the world over and it seems unlikely that triangular theatres were widely, if ever, constructed in India.

Consecration of the Theatre

Consecration of the newly constructed playhouse was obviously important, for an entire chapter of the *Nāṭyaśāstra* is devoted to ritual prescriptions. Bharata says, "Offering worship to the gods of the stage is as meritorious as a (Vedic) sacrifice." He continues:

> If the stage is properly consecrated it will bring good luck to the king (literally, the master) and to people, young and old of the city as well as of the country. But when the auditorium is not consecrated in proper manner it will be indifferently held by gods, and there will be an end of the dramatic spectacle, and it will likewise bring evil to the king. He who willfully transgresses these rules . . . and practices . . . will soon sustain loss and will be reborn as an animal of lower order. (Ghosh 1967, 43-44)

Staging

In chapter fourteen, Bharata says that the stage was divided into different zones and that by walking around from one zone to another, *parikramaṇa,* the actor was understood to have moved from one place of action to another. That this convention was widely accepted may be seen by the ample references that are made to *parikramaṇa* in the extant plays. It is not clear if different parts of the stage were conventionally used over and over again to symbolize the same place, such as a garden or a room in a palace, or whether the whole stage was regarded as a neutral acting area, which did not take on an iconic significance until the actor specified the locale through speeches or actions. Nor may we totally disregard the possibility that scene pieces were employed to designate locales. As mentioned, Bharata classifies, among other things, mountains, temples, caves, aerial vehicles, and houses as stage properties. It may be that these were used in a conventional manner to designate places, as the medieval theatres of Europe designated various locales by employing scene pieces, for the medieval stage was regarded as a neutral acting area *(platea)* until the actors identified themselves with reference to the scenic locales.

Bharata further elaborates on the symbolic use of the stage by suggesting that persons who enter the stage first are regarded as being inside a place or zone and those who enter later are thought to be outside that particular zone and must pass through some barrier, real or imaginary, before they too are regarded as inside the same space. Also in chapter fourteen, Bharata hints that the stage doors may have been used in a conventional manner. Perhaps one was reserved exclusively for entrances and the other for exits. However, it is possible that they may have had a specific symbolic significance, much like that claimed for the doors of the *skene* houses in the ancient Greek theatres.

Dramatic Composition

The casual reader who examines a Sanskrit play in translation for the first time is likely to be somewhat puzzled by what he finds there, unless he has a reader's guide. In this section I will focus on some of he general characteristics of dramatic composition, including lplot, time place, character, and the purpose of drama, as well as some special factors, such as brief examination of fate, duty, illusion, nature, and the language of the play.

Opening Verse and Prologue

Just prior to the main body of *Śākuntala*, and for that matter most Sanskrit plays, there is a short benediction and a prologue. The benediction is composed in verse to honor a deity, Brahman, king, or the like. The prologue usually praises the audience, announces the name of the play and its author, and gives details regarding the content of the work. Obviously the benediction and the prologue serve both a ritual and a practical function. The benediction adds a touch of formality and sanctity necessary to insure that the evil eye is not cast on the proceedings. The information that the stage manager imparts takes the place of some of the essential information found in modern theatre programs. It usually sets the tone for the opening scene and thus is vital to the work. It serves as a bridge between the world of the audience and the world of the play. By having a more important member of the acting company, and not one of the characters of the play, address the audience directly, as a storyteller might do, the two worlds are linked, making the theatre event special and distinct from everyday life.

Preliminaries

According to the *Nāṭyaśāstra,* the benediction and prologue are features of a much more extensive set of introductory items. It may be that the procedures outlined in the *Nāṭyaśāstra* were no longer in vogue when many of the important plays were composed or that some authors took great liberty with the rules and composed prologues according to their own needs. Whatever the case, chapter five lists eighteen different stages of preliminaries *(pūrvaraṅga,* which literally means "before the stage"). The first nine were to be executed on the stage behind a curtain. Each had as its objective to gain the favor of some deity or other. The nine steps are:

1. Arrangement of musical instruments.
2. The singers enter and take their seats.
3. Beginning of vocal exercises.
4. Adjustment of musical instruments.
5. Rehearsal of the different styles of playing.
6. Adjustment of the stringed instruments.
7. The dancers enter and rehearse.
8. The drums and strings are played in unison.
9. Complicated rhythmic passages are practiced.

After a curtain was removed the remaining preliminary items were executed.

10. Songs are sung in praise of the gods.
11. The stage manager raises the banner pole of Indra.

According to the *Nāṭyaśāstra,* this "is considered by some to be the beginning (of the performance)" (Ghosh 1967, 79).

12. The stage manager walks around the stage and offers praise to the guardian deities of the stage.
13. The benediction is recited.
14. The meaning of this item is unclear in the text. Keith believes that it was the recitation by the stage manager of a verse in honor of the god, king, or Brahman whose festival was being celebrated (Keith 1924, 339-340).
15. Keith also explains this item as "the beginning of the dramatic action *(abhinaya),* the Sutradhara reciting another verse and bowing before the banner of Indra" (339-340).
16. Patterns of movement are performed depicting the erotic and then the furious sentiment. [Perhaps this indicates the performance of dance movement which are predominately female *(lāsya)* and male *(tāṇḍva)* in character.]
17. Then a conversation ensues between the stage manager, the clown, and the attendant which is referred to as the three men's talk.
18. Finally, the stage manager announces the content of the drama and appeals for its success, after which he makes an exit and brings the preliminaries to a close.

Objectives

It would seem that several objectives were served by the lengthy preliminaries. First, they provided the musicians with an opportunity to tune their instruments and warm up before the show began. For the dancers it offered an opportunity to practice their dances and gestures. Second, once the curtain was removed the formal songs and dances in praise of the deities were sung in accordance with the general ritual character of the event. Third, the stage manager could offer praise to the banner pole of Indra, which symbolically protected the actors from harm. He begged protection from various gods and thus sanctified the acting area. Fourth, speech began to assume greater importance than physical action and music in the events, and the preliminaries blend into the prologue which serves as the final bridge into the beginning of the drama proper. The *Nāṭyaśāstra* even makes provisions for truncating the preliminaries and also expanding them, which suggests that the author was aware of a need to make compromises to suit any occasion.

Types of Plays

The *Nāṭyaśāstra* identifies ten major types of plays: *nāṭaka, prakaraṇa, aṅka, vyāyoga, bhāṇa, saṃvakāra, vīthī, prahasana, ḍima,* and *īhāmṛga*. This list is apparently not exhaustive, for other ancient writers mention the existence of additional dramatic types. Ten so-called minor *(uparūpaka)* types of drama are discussed in the ancient literature, although the *Nāṭyaśāstra* mentions only one of these in passing. It would seem that there was a wide range of dramatic types within which a writer might choose to compose a play. Yet the *Nāṭyaśāstra* chooses to focus almost exclusively on two of the ten types—the *nāṭaka* and *prakaraṇa*. A number of *nāṭaka* and only two *prakaraṇa* survive to aid us in understanding the differences between these two important dramatic types. Among the extant *nāṭaka* are three of the finest examples of Sanskrit drama—Bhavabhūti's *The Latter History of Rāma,* Bhāsa's *The Vision of Vāsāvadattā,* and Kālidāsa's *Śākuntala*.

The chief characteristics of the *nāṭaka* are that it has as its subject matter a well-known story concerning the exploits of a hero who is either a royal sage or a king; the dominant sentiment of the work should be either love or heroism; and the extant *nāṭaka* normally have no fewer than five and no more than seven acts each. In contrast the *prakaraṇa,* of which *The Little Clay Cart* is the finest example, has a story which is invented by the author, a hero who is either a Brahman, a minister, or a merchant, a heroine who is a courtesan, and the sentiment of love predominates. The *prakaraṇa* should have no less than five and no more than ten acts.

* * * *

Composition

The smallest unit of composition of any play is an act *(aṅka)*. The *Nāṭyaśāstra* lays down numerous rules which govern the composition of acts of plays (Ghosh 1967, 344). For example, an act should portray the change in the hero's basic situation and thus cause the plot to develop. It is made up of a series of incidents directly related to the exploits of the hero, heroine, or a person of similar importance and not to the minor characters. Certain events should not be depicted, such as the pronouncement of a curse, a marriage ceremony, a battle, loss of a kingdom, or death. The *Nāṭyaśāstra* goes on to say that "a wise playwright should not put in (too) many events in one act" (Ghosh 1967, 356).

By putting restrictions on the events that may be depicted in the drama, the rulemakers forced the playwrights to focus on scenes which at first glance seem inconsequential or even trivial. The first three acts of *Śākuntala* are a case in point. In Act I the king meets and instantly falls in love with Śākuntala, learns that she is a suitable match for him and determines to win her for his bride. In Act II he reasserts his determination to remain at the hermitage in order to wed Śākuntala, despite pressing business at the palace. In Act III, the high point of this part of the play, the king secretly overhears Śākuntala imply that she loves him. By the end of the act we assume that they will be married, although she makes no promises to him. In contrast the events that happen between Act III and Act IV are of major consequence—*Śākuntala* and Duṣyanta are married by mutual consent; she becomes pregnant; he departs for the city and her foster father returns to the hermitage, approves of her marriage and determines to send her to the palace where he assumes she will become queen. In the first three acts the playwright has chosen to focus on seemingly minor incidents in the lives of the chief characters and only mentions the major events in the prologue to Act IV. In Sanskrit drama the plot usually stresses the means to the end rather than the end itself. In *Śākuntala* Kālidāsa is more interested in showing the subtle and delicate human emotions of the central characters.

Plot

According to the *Nāṭyaśāstra* the principal objective of the plot is to show the hero struggling for and finally attaining the object of his desire. Subsidiary incidents may contribute to this aim but should not divert attention from it. The realization of the goal relates to the three ends of Hindu life—duty, pleasure, and wealth.

The plot or subject matter is regarded as the "body" of the drama (Ghosh 1967, 355). The anthropomorphic analogy is extended throughout the discussion of this subject. To begin with there are five stages in the development of the action. The first or "beginning" stage refers to the planting of a seed or germ—the desire on the part of the hero to obtain his objective. In *Śākuntala* the king's objective is to wed Śākuntala and to get an heir to his throne. Both of these goals are clearly implied in the excerpt quoted from Act I. Second, there is a determined effort to achieve the object of desire. Duṣyanta's determination in Act II clearly exemplifies his desire to take Śākuntala as his wife. Third, there is hope of success. In Act III Duṣyanta is confident of winning Śākuntala, although she does not commit herself to fulfilling his wishes. Fourth, success is assured if only a difficulty may be overcome. In Acts IV, V, and VI Duṣyanta's original desires are thwarted and delayed. In Act IV we learn that Śākuntala is cursed to be forgotten by the king. In Act V, as a result of the curse, he does not recognize her when she comes to court and in her humiliation she is whisked away by a heavenly nymph to an undisclosed sanctuary. In Act VI, when the king's keepsake ring that he gave Śākuntala in the hermitage is found, his memory is immediately restored and he bemoans the loss of his beloved. The fifth stage in the development of the action occurs when the hero attains the object of his desire. In the final act of *Śākuntala*, Duṣyanta is reunited with Śākuntala and discovers that he is the father of a fine son and heir to the throne.

Corresponding to the five stages of the action are five elements. The first and principal element is the seed. After the seed is planted the incidents expand like a drop of oil on water (Ghosh 1967, 355 and 381). It is this step which sustains the continuity of the action throughout the play. Step three is the intervention of the major subplot and step four is that of the minor subplot. Step five is the denouement in which one of the three objectives of human existence is achieved.

Resulting from the conjunctions of the five stages with the five elements are junctures or links in the plot. The *Nāṭyaśāstra* makes clear that these do not necessarily coincide with a single act but may well overlap several acts. The junctures are: (1) opening, (2) progression, (3) development, (4) pause, or a stage in which the hero deliberates the seed, his desire, and (5) conclusion. The last juncture corresponds to the state of release from pleasure and pain by union of the individual soul with the divine soul. Each of the segments has corresponding limbs without which the play is said to be "unfit for production" (Ghosh 1967, 386). Details regarding the manner in which these are to be employed are wanting in the text. In general it is fair to say that each limb relates to some action or emotion of the hero in each of the stages of the junctures.

The profusion of details regarding the plot of a *nāṭakā* or *prakaraṇa* may seem unnecessarily confusing. But in essence the approach is simple. Sanskrit plays of these two major types were to be created with a definite end in mind. This end was to satisfy the fulfillment of a desire of a hero or chief character in the piece and, by association, his counterpart, the heroine. The play was to end happily, as one of the limbs of the concluding junctures implies. There is no place here for plays which raise controversies that cannot be resolved or that leave the mind disturbed at the end. The process by which the hero attains the object of his desire is steady, starting gradually and progressing rapidly to the conclusion. Unnecessary diversions have no place in Sanskrit plays. Only those diversions which relate in some way to the principal plot are admitted. Rules for the formation of the plot are far more specific in the *Nāṭyaśāstra* than those outlined by Aristotle for Greek drama. This specificity might well have resulted in works which were cold, calculated, and uninteresting, but for nearly a thousand years writers came to grips with the rules and, in many cases, managed to write original and sometimes brilliant theatre fare.

Time

The *Nāṭyaśāstra* clearly states that the events of an act should take place in the course of a single day. But obviously some events take longer than twenty-four hours to complete and are of such magnitude (battles, for example) that they demand accommodation be made for them in the drama. The Sanskrit playwright was given considerable latitude to digress; through a clever device, the introductory scene. Minor or secondary characters, for example, maidservants or the clown, may report the events which have intervened between the previous act and the events that are about to take place, and in this way compress time and avoid depicting subjects which are taboo and, at the same time, make use of them for furthering the dramatic action (Ghosh 1967, 357-58).

Time is often compressed radically within an act. An act may begin at sunrise and end at sunset. Also actions which would normally take a great deal of time to complete in real life, such as the plaiting of a garland, are accomplished in record time on stage. Coupled with the conventional changes of locales, both time and space assumed broad theatrical significance.

Place

Unlike the rules that were laid down for neoclassical French and English drama, rules for establishing and changing locales in a Sanskrit play were quite flexible and conventionalized. An act might begin in one spot and simply by performing a symbolic movement, such as walking around, the actors indicated that they had moved to another locale. *Śākuntala* contains numerous transitions of locale. At the beginning of the play, the king and his charioteer are traveling in the sky on a magic chariot in pursuit of a deer. The description of the landscape below during the chase, and perhaps some conventional movements, help the spectators to imagine the rapid change of place. Suddenly they halt outside the hermitage compound, where the king agrees not to kill the helpless creature which has taken refuge within. It is presumed that the spectators have to imagine the boundaries of the compound. This section of the scene ends as the king commands the charioteer to enter the hermitage. As the act proceeds the locale becomes even more specific until finally we end with the king in a particular spot in the hermitage garden engaged in conversation with the friends of Śākuntala. So we have moved from a broad generalized locale in the air to a very specific locale on the hermitage, all in the space of one short act. Sanskrit playwrights were fond of using this convention of changing locales to further the aim of the play. As we have noted earlier the stage seems to have been regarded as a neutral acting area and given symbolic significance through dialogue and conventional actions—the ancient equivalent to the focusing techniques used in cinema.

Character

In the Sanskrit plays, most characters are types: the just, powerful, handsome, benevolent hero and his equivalent, the beautiful and virtuous heroine; the witty, proverbially hungry clown-companion; trusted and loyal ministers; charming and witty servants; and so on. In *The Little Clay Cart,* there are a host of rogues, wags, a foolish villain, an intimidated judge, a doting mother, a reformed gambler turned Buddhist monk, a thief turned patriotic reformer—all somewhat unusual in the cadre of characters of the major plays. Character portrayal in Sanskrit drama is not based on realism but on the presentation of universal characteristics, emphasizing the typical. This sets Sanskrit theatre apart from much of Western theatre, in which characters are prized for their individuality of spirit and uniqueness of personality, such as Hamlet, Macbeth, Oedipus, or Hedda Gabler.

The tradition of using character types in the theatre may be understood within the framework of three traditions: the literary, the social, and the theatrical. The literary tradition stems principally from the epics, such as the *Rāmāyaṇa* and *Mahābhārata,* and from books of law and those dealing with social behavior. In the epics, characters such as Rāma and Arjuna are idealized. They exhibit certain superhuman character traits. Indeed Rāma is the god Viṣṇu reincarnated in the image of a man. Their regard for propriety and strict codes of behavior in given situations easily marks them as idealized heroic types. They are more alike than they are dissimilar. According to the law books and codes of social behavior, men of certain rank and station in life were expected to behave in a particular way; to do otherwise would be improper, unthinkable. The ritual of social behavior seems to have stemmed from the original separation of men into occupation groups—priests, warriors, tradesmen, and peasants. Each group had particular duties to fulfill and a particular behavior pattern was expected of men in these categories. Perhaps the priests had the most rigid taboos surrounding their daily existence and, as such, refrained from certain social interactions and participated in sacred acts which would be unthinkable for those at the other levels of society.

The broad patterns of behavior are clearly seen in the drama. Despite the caricature of the clown-priest who is the sidekick of the hero, he is always treated with the utmost respect by all those who owe deference to him. The duty of the warrior-hero, such as Duṣyanta, is to protect the Brahmans so that they may make sacrifices to the gods, acts which symbolically sustain the life and power of the heavenly beings. The gods are weakened when offerings cannot be made to them by the Brahmans and thus evil is permitted to gain the upper hand on earth. The social order is sustained when members of every caste fulfill their correct social function. To do otherwise would weaken the social order and wreck the system.

One may well ask, "How does this system permit creativity and individual expression?" In many instances the playwrights have succumbed to slavish perpetuation of the rules of the *Nāṭyaśāstra*, producing lifeless characters. But a number of characters are subtle, fresh, and very much alive. We might even describe them as three dimensional. Śākuntala is one of those. When we first meet her she is a child of nature. Her swelling bosom proclaims that she is a maiden on the verge of blossoming into a charming young woman.

It is this sweetness and innocence that initially attracts Duṣyanta to her. She exchanges no words with her suitor during their first encounter, yet her glances give him hope that she is equally attracted to him. In Act III in the flower arbor she freely and playfully confides to her friends that she is attracted to the king and yet she shyly withdraws into silence when he comes out of hiding and protests his love for her. Here she is growing more daring and on the verge of maturing into a woman. The end of the act is a virtual repeat of the end of Act I when she delays her departure in order to steal a glimpse of the king, confirming that she is attracted to him. In Act IV, generally regarded as the best act of the play by Indian and Western critics alike and some say the most beautiful act of all Sanskrit dramatic literature, Śākuntala must part from all that she loves in the forest hermitage—the plants, deer, her childhood friends, her foster father. She must venture out into the world, to the sophisticated court where she expects her husband to receive her as his wife and queen. The pathos of this act of parting is a high point of the play and offers the spectators an insight into another side of this innocent, trusting girl. When she is rejected in Act V, her retorts and tears give us yet another side of her. In Act VI she is reported to be emaciated with grief, longing for her beloved Duṣyanta who unwittingly has deceived her. When we meet her again in the final act, she is mature, the mother of a handsome and proud little son. Her tenderness and humility are still intact. But she has made a transition from that of a playful, innocent forest dweller to a mature woman, devoted to her husband and their child.

In the personality of Śākuntala, Kālidāsa has painted one of the most striking portrayals of womanhood in all of Indian literature. Certainly he has conformed to the rules of character development outlined in the *Nāṭyaśāstra*, but the character of Śākuntala is imaginatively and creatively constructed by a master, of that there may be no doubt.

Purpose of Drama

Given the multitude of theatrical conventions and character types it is doubtful that Sanskrit theatre was meant to be a reflection of life in ancient India. Rather it was a model of human behavior and in this it served a very definite educational function. To us today there is a great deal to be learned about Indian civilization by examining the plays of this culture. It gives us a perspective of the pervading views of Hindu philosophy and religion, and to some degree those of Buddhism. It tells us a great deal about human behavior and social interaction. The values and mores of the culture are encapsulated in the plays and should be carefully scrutinized for what they teach us about ancient Indian life, and even for what they say about the continuity between ancient and modern life.

When approaching the plays it is well to remember that the wider the students read the better the plays will fit into their cultural framework. A quick reading cannot do justice to intricacies of customs, philosophies, and behavior radically different from our own. Indian drama, like Indian literature in general, has many complicated and subtle nuances. Probing the poetry, epics, and law books of ancient India will inevitably lead to a more comprehensive grasp of the place of drama within the whole fabric of the social setting.

Fate, Duty, Illusion, and Nature

A word or two must be said about several important features of the plays which characterize them as uniquely Indian. In the plays, people are but part of a larger universal order. An individual plays an important part, but when he or she neglects some duty the system is threatened. In *Śākuntala* all seems well at the end of Act III. Śākuntala and Duṣyanta seem destined to lead a long and happy life together. And yet Śākuntala makes one small error. In the prologue to Act IV, her friends report that she has been so engrossed in thinking of her absent lover that she neglects to offer the proper respects to a hotheaded sage who comes as a guest to the hermitage. He curses her to forget the person of whom she is thinking. Not realizing the consequences of her oversight, Śākuntala is doomed to spend many long years away from her beloved.

Fate thus plays a significant role in *Śākuntala* and in most Sanskrit plays. Usually in the opening act we are specifically told, or at least it is alluded to, that the object of desire will be achieved at the end of the play. During the path to the goal, however, the hero and the heroine usually experience some setback, often separation. This tests their love and devotion to each other. And after a period of suffering, the system returns to normal, the wheel of fortune completes its revolution. In Act I of *Śākuntala,* Duṣyanta is blessed by the hermits to have a son who is destined to be a universal ruler. Often the prophesy of success is foretold in this manner.

The characters then are ruled by duty—duty to their people, to their caste obligations, to their social obligations. By neglecting her duty to show utmost respect to a prominent sage, Śākuntala is fated to be punished. The social contract is so great that the slightest infringement of it will bring about serious consequences. The plays seem to have been written to demonstrate the virtue of performing one's duty properly.

The characters are often beset by maya, or illusion. Like the concept of duty, this term is a great deal more complicated to explain than we may allow space for here. However, on a very simple level, maya refers to the notion that the temporal world seems to be permanent, and yet we know that it is constantly changing. The tree that was a sapling yesterday will be a powerful banyan tomorrow and is eventually destined to die or be destroyed. Nothing ever remains the same. Nothing is permanent. People, then, are caught in the eternal web of illusion, thinking they are permanent, when in fact they are constantly changing.

There is a charming illustration of the effects of maya in Act I of *Śākuntala,* when the king chases a bee away from Sakuntala's face. The bee is reported to treat Śākuntala as though she were a flower. The king wishes he were the bee that he might taste the sweet nectar of her loveliness. In Act VI, a picture of the incident is shown to the king and he immediately thinks that the incident depicted in the painting is just happening. He addresses the painted bee as though it were a real bee actually attacking Śākuntala. His companion points out that this is only a picture and that he has temporarily lost his sense of reality. Indeed, the king suffers from maya. He also does so when he forgets Śākuntala as a result of the curse, but with far more serious consequences. The plays then are a subtle example of the workings of maya in the world. The only absolute reality, according to Hindu philosophy, is God who is permanent and who dreams the world. We are but figures in God's dreams.

Finally a word must be said about nature. In *Śākuntala* nature plays a very important part. The hermitage is vividly portrayed. Śākuntala's relationships with the plants, which are as much alive to her as her beloved animals and birds and the people of this idyllic spot, suggest that humans and nature are one and inseparable. It is no accident that Duṣyanta meets Śākuntala and is reunited with her in a hermitage. Śākuntala's haven in the final act is none other than the home of the mother and father of the gods. Court life, city life in general, seems a great deal more contrived and artificial in the plays than does the life in the pleasure gardens of the palace or the hermitages of the forest. Sanskrit theatre paints a vivid picture of the intimate relationship between people and nature, one that places it firmly within the great traditions of dramatic literature of other Asian cultures.

Language

The example of Kālidāsa's *Śākuntala* is typical of Sanskrit plays, in general. Verses are mixed with prose passages. Critics have acknowledged that some of the most beautiful passages of Sanskrit literature are found in the drama, and several passages of *Śākuntala* and *The Little Clay Cart* serve as excellent examples.

It is impossible for a layman to understand much about the subtlety of the Sanskrit language from an English translation, however. Sanskrit is a telegraphic language, full of innuendo. Compressed into a short passage are a multitude of ideas and meanings which are not always possible to translate into another language. Translations cannot distinguish between the levels of Sanskrit spoken by the different characters, nor may we see the fine distinctions between the Prakrits spoken by the heroine and the lofty Sanskrit of the hero. Translations are not capable of rendering the subtle differences among the various regional languages and dialects that must have made Sanskrit theatre a great pleasure and an intellectually challenging art to the ears of ancient listeners.

Aesthetics

The *Nāṭyaśāstra,* outlines a theory of theatre performance that is an original contribution to the field of aesthetics and a major distinguishing feature that separates Indian theatre from all other forms of world theatre. Initially it was applied to theatre performance and subsequently it was borrowed and applied to dramatic literature and expanded to include other forms of Sanskrit literature and other performing arts, as well. Although seemingly simple in its basic outline, it is deceptively so, for critics from a very early date up to the present age have wrestled with the precise meaning of the major terms of the theory and the import of their application.

The theory of *rasa,* or sentiment, attempts to explain how a spectator perceives the performance of a play. This can only be accomplished when he or she interprets what is seen, as well as what is heard and responds to both stimuli as one composite whole. As we have seen earlier, the *Nāṭyaśāstra*, emphasizes that a performance may best be understood in its proper context by those who are learned, wise, skilled in the arts, and sensitive. We have also noted that theatre appears to have been designed for the elite few rather than for the masses. At one point in the text, the aesthetic experience is equated with that of tasting and eating fine foods. Indeed, some scholars have translated *rasa* as "taste." Let us examine this analogy for a moment.

In France haute cuisine is regarded as a high point of cultural experience, engaged in by experts. The permutation and combination of delicately flavored dishes sensitively presented in an appropriate atmosphere is said to delight expert palates, gently leading them through a gastronomical garden of delights to the ultimate conclusion of a meal. The experience is tantamount to a state of exaltation. There are two separate aspects of this aesthetic journey which are intertwined and yet may be viewed separately—the meal itself and the context in which it is served. On the one hand, dishes are chosen to complement and contrast with each other, each dish is prepared with the correct ingredients, combined with care by an expert chef, and presented in a manner which is appealing and conventional. On the other hand, major contributing factors to the total experience are the environment, mood of the participant, time, and place. All these conspire to influence the outcome of the gastronomical event.

The meal is savored by a connoisseur who recognizes the difference between haute cuisine prepared by an expert chef, served in a quiet dining room of a four star restaurant on a quiet summer evening with friends, and meat and potatoes dished up by a harried cook in a steamy kitchen on a Saturday night.

If we equate the theatre experience described in the *Nāṭyaśāstra* to that of an expert dining on haute cuisine, we may more easily follow the basic aspects of the theory. The spectators are like the connoisseurs of cooked food (called bhaktas, literally the "devotees" of god) and the performers may be equated both with the chef and the meal. The actors, musicians, dancers, playwrights, and other functionaries are all expected to prepare and train themselves according to the rules and conventions, for the moment when they will offer their lavish feast for the spectators to savor. Like the priests who officiate at the temple sacrifices, the performers are obliged to perform with seriousness and care so that they do not cause the evil eye to be cast on the whole affair, spoiling and polluting the proceedings and causing the judicious to find fault. In this respect, the theatre event was no ordinary event but a special occasion. It was expected that it would be executed on a very high plane, in a similar vein to that of a religious sacrifice.

The *Nāṭyaśāstra*, divides human experience into eight basic sentiments: erotic, comic, pathetic, furious, heroic, terrible, odious, and marvelous. It is these which the spectators savor in various permutations and combinations as they observe the work in performance. And yet to give focus to the experience, just as a meal has a dominant flavor, one sentiment should dominate all others in the total context of the performance. The most common dominant sentiment of the extant plays is love. And yet all the plays, even the short one-act works, employ varieties of feelings and emotions which provide the needed variety and texture.

Corresponding to the eight sentiments are eight emotions or feelings (*bhāva*). They are expressed by the actor in the process of laying his or her bouquet of experiences derived from the play before the spectators. They are correspondingly love, mirth, anger, sorrow, energy, terror, disgust, and astonishment.

In addition there are thirty-three transitory feelings: despondency, weakness, apprehension, envy, intoxication, weariness, indolence, depression, anxiety, distraction, recollection, contentment, shame, inconstancy, joy, agitation, stupor, arrogance, despair, impatience, sleep, epilepsy, dreaming, awakening, indignation, dissimulation, cruelty, assurance, sickness, insanity, death, fright, and deliberation. And eight *sāttvika* states (although the exact meaning of the word has not yet been satisfactorily translated into English, we have used both "emotion" and "feeling," as a definition elsewhere in this chapter) which are paralysis, perspiration, horripilation, change of voice, trembling, change of color, weeping, and fainting. This makes forty-nine different feelings in all. However, the *Nāṭyaśāstra,* clearly states that "just as a king is superior to other men, and the preceptor is superior to his disciples, so the *bhāva* are superior to the transitory and *sāttvika* states" (Ghosh 1967, 122).

The emotions are perceived by the spectators in the actors and evoke a corresponding sentiment in the spectators. At one point in the *Nāṭyaśāstra,* the emotions and sentiments seem to refer to the same thing. Indeed it is somewhat difficult to distinguish between them.

It is said that every play has a dominant emotion just as it has a dominant sentiment. The dominant emotion may be easily determined by the expert spectator (or reader) from among various transitory feelings, for a play performed with only one emotion would be like consuming a meal in which all the food tasted exactly the same.

It is said an actor produces an emotion and causes a sentiment to be perceived through the union of the determinants, the consequents, and the transitory feelings. Perhaps this may be best understood by citing a simple illustration. Rāma, the epic hero of the *Rāmāyaṇa,* acts as though he sees the beauty of the forest (determinant) which causes his eyes to widen and his shoulders to tremble with pleasure (consequent). He stands stunned and overjoyed (transitory feeling). Thus the actor portrays this emotion of astonishment. If the actor has portrayed the emotion correctly and with feeling, the experienced and sensitive spectator will savor the sentiment of wonder.

Within the context of an act of a play this example may be but a tiny part of a much broader pattern in which wonder becomes a minor factor. Indeed, the act of the play from which the illustration was taken emphasizes the dominant emotion of love. And the entire work lays particular stress on astonishment *(vismaya).*

According to the dramatic theory, the play is always the touchstone of the actor's portrayal, and the character will dictate to what extent the actor should depict the emotions of the play and the general rules governing his preparation. It would seem that the system just outlined leaves virtually nothing to chance. In actuality a great deal of latitude is provided. Obviously not all actors are equally capable of portraying emotions. Some are brilliant technicians but lack a well-developed ability to portray emotions. Others are highly emotional but lack the skill and technique to depict feelings within the framework of a set of rules and conventions. There are others who are both brilliant in their craft and artists. If the *Nāṭyaśāstra* served as the actor's text, it is safe to assume that ancient Indian performers possessed at least an adequate set of conventional techniques, even if they lacked the necessary emotional base to lift the performance from the ordinary to the sublime. In this regard, the theory of sentiment is a practical guide to a performer in developing his or her craft, as well as being a useful tool in realizing the maximum potential of the art.

Decline of Sanskrit Theatre

The decline of Sanskrit drama cannot be attributed to any one cause but to a combination of factors. It is certain that it did not die suddenly.

The signs of decay are evident as early as the ninth century, and it is generally believed that the tenth century marks the end of the Sanskrit theatre as a creative force in Indian theatre. In practice its influence is still being felt, as we shall see at various points in the remainder of this book. A major factor in its decline may be the successive invasions of Mahmud of Ghazni and successive waves of invasions of other plunderers from kingdoms to the west of India. At this time India was divided into small princely states, each vying with the other for supremacy. The armies of Ghazni were well organized and tactically superior to those of the kings of the subcontinent, and so the old court life of Hindu India began to crumble under the pressure of warrior lords that plundered, looted, sacked, and razed cities and temples with only one objective in mind—to confiscate the wealth of India.

By one scenario, during these harsh times courts and temples could ill afford the luxury of supporting the performance of a Sanskrit play, so this put the actors on foot trying to eke out a living by touring from village to village. But there too they met with little success, for Sanskrit was not a language that was widely spoken by the people of the rural areas. Except for the Brahman priests very few people outside the courts could understand the texts of the plays. The players had very few options—either they reverted to performing in a regional language of the locality or they gave up their hereditary occupation. The practice of performing Sanskrit plays was doomed to disappear over a period of centuries, in all but a few isolated pockets around the country.

Another reason for the decline of Sanskrit drama lay in the trends in composition which appear as early as the plays of Bhāsa. Greater and greater emphasis was placed on the poetry and less and less on the dramatic action and the dialogue. Kālidāsa and later poets, especially Bhavabhūti, laid heavy stress on the lyrics, often at the expense of the progress of the story. This made the works seem more like long poems than plays. Strict adherence to the rules of the *Nāṭyaśāstra* also may have contributed to the downfall of this form of aesthetic expression.

It is rare that talented individuals had the imagination and the daring to create plays outside the strict rules. True, the rules often served merely as guidelines but when critics were harsh to judge the merits of a play according to the rules, the playwrights must have succumbed to authority and composed according to form to win approval.

Thus, a form of theatre which represents a high water mark in the history of world theatre began to pass from the scene.

Works Cited

Bharata. 1967, 1961. *Nāṭyaśāstra.* Trans. and ed. Manmohan Ghosh. 2nd rev. ed. Vol. 1. Calcutta: Manisha Granthalaya, 1967; and Vol. 2. Calcutta: Asiatic Society, 1961.

Kalidasa. 1964. *Sakuntala, or the Recovered Ring.* Trans. A. Hjalmar Edgren. In Wells 1964. Pp. 195-281.

Keith, A. Berriedale. 1924. *The Sanskrit Drama.* London: Oxford University Press. 1971.

Rangaçharya, Adya. 1971. *The Indian Theatre.* New Delhi: National Book Trust.

Śudraka. 1964. *The Little Clay Cart.* Trans. Revilo Pendleton Oliver. In Wells 1964. Pp. 39-193.

Wells, Henry W. Ed. 1964. *Six Sanskrit Plays.* Bombay: Asia Publishing House.

Additional Readings

Baumer, Rachel Van M., and James R. Brandon. Eds. 1981. *Sanskrit Drama in Performance.* Honolulu: The University Press of Hawaii.

Bhat, G. K. 1959. *The Viduśaka.* Ahmedabad: The New Order Book Co.

Chaudhury, Pravas Jivan. 1952. "The Theory of *Rasa.*" *Journal of Aesthetics and Art Criticism 11,* 2:147-150.

Coulson, Michael. Trans. 1981. *Three Sanskrit Plays.* New York: Penguin.

Dace, Wallace. 1963. "The Concept of Rasa in Sanskrit Dramatic Theory." *Educational Theatre Journal* 15:249-254.

Dhanaṃjaya. 1962. *The Daśarūpa: A Treatise on Hindu Dramaturgy.* Trans. George C. 0. Haas. Delhi: Motilal Banarsidass.

Gerow, Edwin. 1977. "Indian Poetics." In *A History of Indian Literature.* Ed. Jan Gonda. Vol. V. Wiesbaden: Otto Harrassowitz. Pp. 217-301.

Kale, Pramod. 1974. *The Theatric Universe.* Bombay: Popular Prakashan.

Levi, Sylvain. 1978. *The Theatre of India. Vols.* I and II. Trans. Narayan Mukherji. Calcutta: Writers Workshop.

Nandikesvara. 1975. *Abhinayadarpanam.* Trans. Manomohan Ghosh 3rd ed.

Calcutta: Manisha Granthalaya.

Raghavan, V. 1958. "The Aesthetics of Ancient Indian Drama." *Indian Literature* 1, 2:67-74. 1967.

———. "Sanskrit Drama: Theory and Performance." *Comparative Drama* 1, 1: 36-48.

Rangacharya, Adya. 1967. *Drama in Sanskrit Literature*. 2nd revised ed. Bombay: Popular Prakashan.

Stoler Miller, Barbara. Ed. 1984. Theater *of Memory: The Plays of Kalidasa*. New York: Columbia University Press.

Swann, Darius L. 1969. "Indian and Greek Drama: Two Definitions." *Comparative Drama* 3, 2:110-120.

Van Buitenen, J. A. B. 1968. *Two Plays of Ancient India*. New York: Columbia University Press.

Vatsyayan, Kapila. 1968. *Classical Indian Dance in Literature and the Arts*. Delhi: Sangeet Natak Akademi.

Wells, Henry W. 1963. The Classical Drama of India: Studies in Its Value for the Literature and Theatre *of the World*. New York: Asia Publishing House.

———. Trans. 1968. *Sanskrit Plays from Epic Sources*. Baroda: Maharaja Sayajirao University of Baroda.

Winternitz, M. 1963.*History of Indian Literature*. Delhi: Motilal Banarsidass.

Zarrilli, Phillip B. 1987. " 'Where the Hand is [. . .].' " *Asian Theatre Journal* 4, 2:205-214.

MRICHCHHAKATIKA

(THE LITTLE CLAY CART)

PRAKARANA

(*A Play in Ten Acts*)

Characters in the Play

SUTRADHARA	Manager of the Play
NATI	Wife of SUTRADHARA
CHARUDATTA	Hero of the Play, a Brahmana merchant who is a citizen of Ujjayini
MAITREYA (V̇IDUSHAKA)	Friend and companion of CHARUDATTA; a Brahmana
VARDHAMANAKA (CHETA)	Servant of CHARUDATTA
SAMVAHAKA (later MENDICANT)	A professional shampooer, who takes to gambling and later becomes a Buddhist monk
RADANIKA	CHARUDATTA's maid-servant
DHUTA	Wife of CHARUDATTA
ROHASENA	Son of CHARUDATTA
VASANTASENA	Heroine of the Play, a courtesan who loves CHARUDATTA
KARNAPURAKA	Servant of VASANTASENA
MADANIKA	VASANTASENA's maid-servant
KUMBHILAKA	Attendant of VASANTASENA
MAID	Attendant of VASANTASENA
VITA	Courtier of VASANTASENA
OLD LADY	Mother of VASANTASENA
BASTARDS	Referred as *Bandhula*, Parasite of VASANTASENA
UMBRELLA BEARER	Attendant of VASANTASENA

SAKARA (SAMSTHANAKA)	Villain of the play who is brother-in-law of the king of Ujjayini
VITA	Friend of SAKARA
STAVARAKA (CHETA)	Servant of SAKARA
MATHURA	Keeper of a gambling den
DARDURAKA	Gambler
DHUTKARA	Gambler (Gamester)
SARVILAKA	A Brahmana-adventurer who commits a robbery for obtaining money to ransom-off the servant-girl MADANIKA whom he loves
ARYAKA	A cowherd and insurgent, later the king of Ujjayini
VIRAKA, CHANDANAKA	Two officers of the city police
JUDGE	Judge of the law-court
SODHANAKA	Beadle of law-court
SHRESHTHIN, KAYASTHA	Two assessors
TWO CHANDALAS	Two executioners (Ahinta and Goha)

Other Characters Mentioned:

CHURNAVRIDDHA	Friend of CHARUDATTA
KING PALAKA	King of Ujjayini, later dethroned and killed by ARYAKA
REBHILA	A merchant of Ujjayini; friend of CHARUDATTA; brother of VASANTASENA

ACT I

Prologue

May the meditation of Siva[1], concentrated on Brahma owing to the intense absorption produced by the absence of any perception, protect you! Of Siva, whose knees are wrapped round with the doubly-tightened fold of the snake, while he sits having formed a knot with his legs in the *Paryanka*[2] posture: the operation of whose sense is so suspended as to cause the cessation of all perception, by reason of the vital airs being confined within; and who, with the eye of truth, is visualizing himself and the Universal Soul as one.

Moreover—

May the neck of the black-throated deity[3], Siva, resembling a dark cloud, protect you! The neck whereon the creeper-like arm of Parvati shines like a streak of lightning.

[*End of the Benediction*]

1. Sanskrit writers generally introduce their works with a benediction, or a salutation to their favourite deity. In Sanskrit drama, this purpose is served by the Nandi, which is enjoined by Bharata as necessary for the removal of obstacles and thereby for the proper completion of the work undertaken. We find playwrights from Sudraka onwards conforming to this rule of Bharata. Bhasa, Sudraka's predecessor, however, often dispenses with the Nandi, probably treating it as the last part of the theatrical preliminaries.
2. Cross-legged. Generally known as *Virasana*, it is one of the five principal postures assumed in the practice of Yoga—sitting with the upper part of the body raised straight up perpendicularly, with one foot placed on the other thigh passing it through the knee-joint of that leg, and then placing the other foot on the other thigh. Siva sat in the *Virasana* posture and the serpent lay doubly coiled round the squatting position of the body.
3. Another name for Siva, after the colour of his throat that changed to a dark-blue in consequence of his having drunk the deadly poison produced at the churning of the ocean for obtaining *Amrita* (ambrosia). The throat is compared to a dark cloud because both relieve torment. The second sloka (verse) written by the poet probably suggests the subject-matter of the plot

SUTRADHARA[4]: Enough of an exertion causes distraction of the curiosity of the audience. Having saluted these honourable gentlemen, I will thus respectfully address them: that here we have undertaken to stage this *Prakarana*[5] called *Mrichchhakatika*[6]. It is stated that the poet, its author—was a famous scholar, Sudraka by name, who was the very best of the twice born Kshatriyas[7], and of unfathomable mental calibre. His gait was like that of an elephant, his eyes resembled those of a *chakora* bird[8], his face vied with the full moon in beauty, and he

---

as required by the canon of dramaturgy. Thus, in the play, the casting of the arm of Gauri around Siva's neck suggests the love between Charudatta and Vasantasena, while the reference is meant to indicate the visit of Vasantasena to Charudatta amidst rain and flashes of lightning. The contrast between white and blue further suggests the dark ways of the world, as exemplified by the conduct of Samsthanaka, and the joys of pure love as in the case of Vasantasena. The Nandi here consists of eight *padas* (*pada* means a line of a sloka); casting the *bija* (seed of the plot) is indirectly suggested.

4. In Sanskrit drama, it is generally the Sutradhara who enters the stage first and initiates the *prastavana* (prologue) after the Nandi path. In *Mrichchhakatika*, the entry of the Sutradhara is not mentioned as a stage direction. Also, the entry of Sutradhara is not placed before the introductory sloka, because it contains a benediction and forms a part of *purvaranga* (literally 'before the play'; conventional preliminaries like the Nandi & Prastavana). It is arguable whether Sutradhara was already on stage when the Nandi was recited. Devendra Raj Ankur's essay included in this volume, *Mrichchhakatika: A Reading*, discusses this point.
5. A category in *dasarupaka* (ten kinds of drama). See Glossary.
6. The title of Sudraka's *Mrichchhakatika* comes from the earthen play-cart made for Rohasena, Charudatta's son, as mentioned in the sixth Act. This incidence certainly serves as the turning point in the development of the plot; thus the significance. For details on the various implications of the 'play-cart', read Devendra Raj Ankur's essay included in this edition.
7. Originally, the word *dwij* (twice born) signified the first three classes of the Hindu society, viz. Brahmanas, Kshatriyas and Vaisyas. It was only later that it came to refer to the Brahmana class exclusively. Here, it is a reference to the most eminent of Kshatriyas.
8. A kind of partridge. Legendary bird in Hindu mythology. The eyes of a *Chakora* (the Greek partridge) are reddish and beautiful, and are therefore commonly used as an epithet to describe eyes of women. Sudraka implies the natural redness to indicate heroism. Here, the meaning is 'eyes free from the blindness of ignorance', i.e., eyes of knowledge.

possessed a handsome body.

Again—

He knew the Rigveda, the Samaveda, Mathematics, the Fine Arts practised by courtesans[9], and the Science of the training and management of Elephants, and he obtained, through Siva's favour, eyes from which the darkness of ignorance had departed. Having seen his son installed on the throne as king, having performed the exalting *Asvamedha*[10] sacrifice, and having enjoyed a life of one hundred years together with ten more days, Śudraka eventually entered the sacred fire.

Moreover—

Passionately fond of war, free from negligence, chief of those who knew the Vedas, rich in penance, and eager to fight with the elephants of his enemies with bare arms—such indeed was King Sudraka.

And in this work of his—

It is related that, there lived, as was said, in the city of Ujjayini[11], Charudatta, a leader of the Brahmanas, who was young and poor, and a courtesan called Vasantasena, lovely like the beauty of *Vasanta*[12], who loved him for his virtues.

Through Vasantasena and Charudatta, King Sudraka has excellently delineated the course of virtuous life based on the pleasures of true love, corrupt legal procedure, the nature of villains, and workings of Destiny.

9. *Vaishiki kala* in Sanskrit, meaning 'belonging to the Vesa' (also, vaishya). It means the arts to be found among or practiced by the courtesans; the elegant arts. The knowledge of the following arts among others was deemed necessary for the courtesan: dance, song, instrument *et al.* See Glossary entry, 'Courtesan.'
10. In ancient India it was the highest ambition of any king enjoying supreme power to perform a horse-sacrifice at least once. One such was performed, and perhaps the last of the kind, by Pushpamitra, the father of Agnimitra, the hero of the Malav. See Glossary.
11. Situated on the Sipra river, Ujjayini's ancient name, Avantpuri, is derived from the capital of the country called Avanti, which formed the eastern part of Malwa. The modern Ujjain is about a mile south of the ancient city.
12. Spring.

(*walking about and observing*) How now! This music-hall of ours is quite empty! Where could the actors have gone? (*reflecting*) Ah! I got it!—

The home of a sonless person is empty; he who doesn't have a real friend finds all time empty; the quarters are empty to a fool; and everything is empty to a poor man.[13]

Well, I have finished the musical concert. By having thus practised music for a long time, my restless eyes, are crackling with pain owing to hunger, like lotus-seeds scorched by the rays of the fierce sun in summer-time. I shall therefore now call my wife and ask her if there is any breakfast available. (*to the audience*) Here have I, sirs, turned into a Prakrit-speaking person, as required by my business and the exigency of the performance too. Alack-a-day![14] By having practised music for a long time my limbs are drooping in hunger, like the parched stalks of lotuses. I shall therefore now go home and ascertain if my wife has arranged for anything. (*walking about and observing*) Here is my home; I shall go in. (*entering and observing*) Heigh-ho![15] How is it that there is quite a new development in our house! The passage is full of a long stream of rice-water, and the floor, which is dark-grey because of the iron saucepans moved over it, is now looking more charming, like a decked-up woman. And my hunger is agonising me the more, as if it were greatly excited by this inviting aroma. Could it be, then, that some buried ancestral treasure has been discovered? Or is it me who in hunger is looking upon the world as consisting of nothing but eatables? Indeed, there could not be any breakfast in our house. Hunger is killing me so as to take away my very life. Here all things are wearing a new aspect today. One woman is pounding unguents, while another is wreathing flowers. (*pondering*) What could this be? Nevermind, I shall call for my wife and ascertain what it all really is. (*looking offstage*) Lady, come here, please.

13. In the translation by M.R. Kale, the word 'shunya' is translated as 'empty' in all references. It may also be translated as: "The home of a sonless person is barren; he who doesn't have a real friend finds all time void; the directions are meaningless to a fool; and everything is nothing to a poor man."
14. Exclamation made to express regret or dismay.
15. Expressing boredom, resignation, or joy. Here it is used in a sense of joy.

NATI (*entering*): Sir, here I am.

SUTRADHARA: Lady, my greetings to you!

NATI: May Your Honour command what duty is to be executed.

SUTRADHARA: Lady, (*Saying—'By having practised music for a long time' et al*) is there anything to eat in our house?

NATI: Sir, there is everything.

SUTRADHARA: What things are there?

NATI: Rice seasoned with *guda*[16], ghee, curds, rice, and all elixir-like things fit to be eaten by you. May the gods thus favour you!

SUTRADHARA: What! Are all these things really available in our house? Or are you joking?

NATI (*to herself*): Well, I'll joke. (*aloud*) Husband, they are available, in the market.

SUTRADHARA (*with anger*): Ah! You hussy, I have been raised aloft and thrown down, like the stone at the end of the long beam of bamboo. Your hopes will be similarly shattered, and you too would end.

NATI: Pardon sir, pardon! I indeed played a joke.

SUTRADHARA: Well, then, what are these preparations that seem to be new here? One lady is pounding unguents, another is wreathing flowers, and this floor here is decked with the offerings of flowers of five colours.

NATI: I have today observed a fast.

SUTRADHARA: What is this fast called?

NATI: It is called *Abhirupa-Pati*[17].

SUTRADHARA: Lady, in this world or in the next?

NATI: Sir, in the next.

SUTRADHARA (*to the audience; angrily*): Look, worthy gentlemen, look! At the expense of *my* food, a good husband is sought for in the next world!

NATI: Be pleased, sir, be pleased! I have observed the fast so that you yourself should be my husband in the next birth.

16. Jaggery.

17. A fast observed to be blessed with a handsome husband.

SUTRADHARA: By whom was this fast recommended to you?

NATI: By Churnavriddha, your own dear friend.

SUTRADHARA (*angrily*): Ah! You son-of-a-slave, Churnavriddha! When, indeed, shall I see you being cut up[18] like the scented, braided hair of a new bride as ordered by the enraged King Palaka!

NATI: May Your Honour be appeased! This fast[19], which is to bear fruit in the next world, is with reference to no one but Your Honour. (*she falls at his feet*)

SUTRADHARA: Lady, get up. Pray, get up. Tell me what is required for this fast.

NATI: We must invite a Brahmana befitting people like us.

SUTRADHARA: If that is so, you may carry on, I too shall invite a Brahmana suitable for people like us.

NATI: As Your Honour commands. (*exit*)

SUTRADHARA (*walking about*): Well, well! Now, how am I to find, in this very prosperous city of Ujjayini, a Brahman befitting us? (*seeing*) Here is this Maitreya, Charudatta's friend, coming in this very direction. Well, I will ask him first. (*to* Maitreya) Sir Maitreya, kindly be the first to dine in our house today.

MAITREYA (*offstage*): Sir, you should invite some other Brahmana. I am engaged at present.

SUTRADHARA: Sir, the food is rich, and there is no rival. Moreover, there will be some *Dakshina*[20] too, for you.

MAITREYA (*offstage*): Sir, I have already refused. Why this persistence on your part now to importune me again and again?

SUTRADHARA (*to the audience*): I am given a refusal by him. Well, I shall invite some other Brahmana.

(*Exit*)

END OF THE PROLOGUE

18. The object of the poet is to suggest that the ordering of the cutting of the hair of a young woman by the king would make his act revolting in the eyes of the people and thereby lead to disaffection.
19. A Brahmana must be fed at the time of breaking a fast, otherwise observing of the fast would be rendered futile.
20. Money and gifts given to Brahmanas on various religious occasions.

(Maitreya *entering with a cloak in his hand*)

MAITREYA (*mumbling*): Or I, too, Maitreya, have to look for invitations from others! Ah! Fortune, you have made a pauper out of me! I, who indeed, when my friend Charudatta was prosperous, used to devour, day and night, upon the sweetmeats, emitting savoury smells at the effusion of breath, and prepared with great care, and I, sitting at the entrance of his *antahpura*[21], surrounded by hundreds of dishes, used repeatedly to touch them with my fingers and put them aside when satisfied, just like a painter surrounded by hundreds of colour pots touching each with his fingers and removing it after use, and who used to remain ruminating like a bull in a public square of the city. I, the very same man, now take my food anywhere at random, owing to his being reduced to poverty, and come here like a domestic pigeon merely to rest at night. Churnavriddha, who is a dear friend of our noble Charudatta, has sent with me this cloak scented with jasmine-flowers, asking that it should be presented to the noble Charudatta after he has finished his daily worship of the gods. Well, then, I will first see Charudatta. *(walking about and observing)* Here comes Charudatta in this very direction, carrying the *Bali*[22] to the household deities after having finished his worshiping of the gods.

(*Enter* Charudatta *as described, and* Radanika)

CHARUDATTA (*looking upwards and sighing dejectedly*):—

On the grounds outside my house, ample offering was made which was formerly eagerly eaten by swans and flocks of doves, on those same grounds now, overgrown with shoots of grass, lies the leftover handful of corn devoured by insects' mouths. (*with these words he walks about slowly and slowly, and sits down*)

VIDUSHAKA: Here is Charudatta. Well, then, I shall now approach him. (*going up to him*) My greetings! May you prosper!

CHARUDATTA: Ah! Here comes Maitreya, our friend for all times. Friend, you are welcome, please be seated.

VIDUSHAKA: As you command. (*sitting*) Friend, your dear friend Churnavriddha has sent this cloak, scented with the blossoms of jasmine, asking me to present it to you when you had finished your

21. Inner room. See Glossary.

22. Ritualistic offering, oblation.

worship of the deities. (*he hands it over;* Charudatta *takes it and remains absorbed in thought*)

VIDUSHAKA: Friend, what are you are thinking about?

CHARUDATTA: Friend,—

Verily, happiness appears splendid after a person has experienced troubles, like the sight of a lamp when there has been pitchy darkness. But when a man is reduced to penury after he has enjoyed luxury, he lives like a dead man, existing only by keeping up his body.

VIDUSHAKA: O Friend, out of death or poverty, what do you prefer?

CHARUDATTA: Friend,—

Out of poverty and death, I prefer death, for death causes short-lived pain, while poverty is unending misery.

VIDUSHAKA: Friend, you need not worry. Even in this condition, you—who' have bestowed his wealth on the needy—look more charming, like the moon of the first day of the lunar month, that remains after it has been drunk up by the gods.

CHARUDATTA: Friend, loss of money is not the cause of my misery. Observe,—

What pains me is that guests avoid coming to my house because it has lost its riches, as the bees fly off to desert, when the thick line of rut on the elephant's cheek totally dries up.

VIDUSHAKA: Friend, these fleeting riches—as insignificant as a morning meal—visit those places only where they are not used, just like the cow-herd's sons in a forest who, afraid of wasps, hide where they would not be bitten by them.

CHARUDATTA: Friend, I am not really feeling any anxiety on account of the loss of my fortune, for riches come and go, following the course of one's luck. But this burns me that people go thoughtless even in their affection towards a person who has lost his support of wealth.

Moreover—

From penury a person passes to shame; being overcome by shame, he loses his spirit; devoid of spirit he is slighted; being slighted, he feels dejected; full of dejection, he comes to be sorry; being smitten with sorrow, he is forsaken by his reason; and destitute of reason, he perishes. Ah! Pennilessness is the abode of all sorts of misfortunes!

VIDUSHAKA: Friend, enough of feeling uneasy by remembering that same wealth which is of as little consequence as a morning meal.

CHARUDATTA: Friend, poverty in the case of a man,—is the abode of anxiety, it is the source of the highest insult; it is a different phase of enmity; it breeds disgust in his friends; and it is the exciter of the hatred of his kinsmen and of the general public. He feels an inclination to retire to the forest; he suffers insults even from his own wife. It is the fire of grief that dwells in his heart; it does not actually burn him, but it pains him continually.

Well, friend, I have made the offering to the household deities. You, too, go and offer a *Bali* to the Holy Mothers at the public square.

VIDUSHAKA: I shan't go.

CHARUDATTA: Why not?

VIDUSHAKA: For, even when thus worshipped, the deities do not show their favour to you. Where is then the use of worshipping them?

CHARUDATTA: Nay, friend, say not so. This is the bounden duty of a house holder.—

Being worshipped by means of physical penance, a devoted mind, praises, and the offerings of oblations, the deities are always pleased with their tranquil-minded devotees. What is the use of questionings about it?

Go, then, and offer the oblation to the Holy Mothers.

VIDUSHAKA: Sir, I won't go. You may order somebody else. For, in my case, a poor Brahmana, everything turns out in a contrary manner: just like the reflection in a mirror, the left appearing to the right and the right to the left. Moreover, at this time of the night there are people on this royal road, courtesans, their parasites and attendants, and men who are royal favourites. I will, therefore, just fall a prey to them, if I come face to face with them, like a mouse with a black snake greedy of frogs. And what will you do sitting here?

CHARUDATTA: All right, wait awhile. I will go through my holy meditation.

(*Offstage*)

Stay, Vasantasena, stay!

(Enter Vasantasena, *being followed by* Vita,[23] Sakara[24] *and* Cheta[25]*)*

VITA: Stay, Vasantasena, stay!—

Why are you, setting aside your tenderness of limbs owing to fear, and plying your feet which are trained by the practice of dancing, running away like a female deer, frightened when being pursued by a hunter, with eyes casting affrighted and tremulous side-glances?

SAKARA: Stay, Vasantasena stay!—

Why do you go, run and scamper and stumble? Be pleased, O girl! You won't die. Stay awhile. Indeed, my poor heart is being burnt by passion, just like a piece of flesh fallen on a heap of burning charcoal.

CHETA: Stay, lady, stay!—

You are running away from me in terror, like a pea-hen in the hot season with her tail fully grown and outstretched. Here is my lord and master bounding (after you) like a young cook in the forest.

VITA: Stay, Vasantasena, stay!—

Why do you run away, quaking like a young plantain tree,[26] your red silken garment trailing with its ends fluttering in the breeze, and dropping off the numerous buds of red lotuses, like a cave of red arsenic, being excavated with a chisel?

SAKARA: Stay, Vasantasena, stay!—

You inflame my love, my passion, my lust, and forcibly deprive me of my sleep when lying on my bed at night; and you are running

23. Companion of a prince, a dissipated young man, or a courtier.
24. (Also Samsthanaka; also referred to as 'Bastard') Brother-in-law of the king, being the brother of his concubine. Sakara ought to be represented with a mixture of pride, folly, and vanity, of low origin, and raised to power only on account of his relation to the king. He is called Sakara because he speaks the Sakari dialect, which abounds in the 'shh' sounds. It will be seen that the speeches of Sakara are full of repetitions, absurd comparisons, and irrelevancies.
25. A servant who assists the master in a love affair. In the play, Vita and Cheta are the pleasure-companions of Sakara.
26. Vasantasena, clad in a red garment and trembling, is befittingly compared to a Kadali plant which is covered with red flowers and can be easily shaken off by the wind.

away stumbling and falling, being agitated by fear, now that you have come into my power, like Kunti into that of Ravana[27].

VITA: O Vasantasena,—

Why do you, surpassing my steps with your own, run away like a female snake overcome with the terror of the eagle? Starting off with speed, I can overtake wind itself, but to overtake you, O fair-limbed one, I don't want to seize you by force.

SAKARA: Friend, friend!—

She is an inciter of the passion of the stealers of coins; she is a fish-eater; a figurante[28]; a flat-nosed maiden; a destroyer of families; an untamable shrew; a casket of love; a courtesan; an inhabitant of the harlot's repository of fine ornaments; a prostitute; and a concubine. I have given her these ten names; and still she desires me not.

VITA:—

Why do you run away, out of fear, while the sides of your cheeks are rubbed against by your earrings tossing to and fro, like a lute hard-struck with their nails by *vitas*, or like a female crane frightened by the rumbling of clouds?

SAKARA:—

Mixing up the sounds made by your numerous jingling ornaments, why do you fly away like Draupadi afraid of Rama[29]? Here I shall carry you off suddenly, as Hanumat did with Subhadra, the sister of Viṣvavasu.[30]

27. There is no connection whatsoever between these two who lived at different mythological times. Such confusion of persons and events is characteristic of the speech of Sakara who thus evinces his ignorance as well as pretension.
28. (Archaic) Pertaining to female actor.
29. Sakara makes blunders of his mythological references, adding further to the tinge of humor.
30. Here again we have funny references to several names that have no connection whatsoever. Visvavasu was a Gandharva king mentioned in the *Mahabharata*, and has no connection with Subhadra. This confusion may be partly due to Sakara's ignorance and carelessness and partly feigned, owing to theatrical needs. For the character of Sakara, like that of the Vidushaka, is meant to stir laughter among the audience. And we know that Sakara is not altogether such a great blockhead as he appears to be; there is a great deal of shrewdness and sagacity in him, as will appear now and then, from Act IX.

CHETA: Admit your caresses to the royal Sakara, and then you will get fish and flesh to eat. In the palace, even dogs feed not on dead bodies.[31]

VITA: Madam, Vasantasena,—

Why do you—wearing the girdle-zone fastened round your sloping hips and shining with lustre like stars and with a face excelling in ruddiness of complexion, powdered red arsenic—run away quickly in fright, looking like the guardian deity of the city?

SAKARA:—

Being hard pursued by us, like a she-jackal by dogs in the forest, you are running away quickly, and with speed, hastily taking away my heart together with its stem.

VASANTASENA: Pallavaka, Pallavaka! Parabhritika, Parabhritika?

SAKARA (*with fear*): Friend, friend! Men, men!

VITA: Fear not, fear not.

VASANTASENA: Madhavika, Madhavika!

VITA (*with a laugh*): You fool! She is looking for her servants!

SAKARA: Friend, friend! I's she calling some woman?

VITA: Surely.

SAKARA: I can kill a hundred of women! I am very brave!

VASANTASENA (*seeing no one around*): Oh! Ill-luck! How now! Even the servants have disappeared! Now I have got to protect myself on my own.

VITA: Seek, seek them out!

SAKARA: Vasantasena, cry out, cry out for your Parabhritika, or your Pallavaka, or the whole of the month of Vasanta. When I am pursuing you, who can save you?—

Can Bhimasena, the son of Jamadagni, or can the ten-necked Ravana, the Son of Kunti save you? Here by seizing you by your massive hair, I am going to play Duhshasana.

Look here, I say, look here!—

My sword is very keen, and your head is right here. Shall I dissect

31. Cheta refers to the dogs of Samsthanaka, so plentifully fed on fish and flesh that they don't care to eat dead bodies.

your head, or shall I kill you? It is futile to run now. Who is destined to die can never live.

VASANTASENA: Sir, I am only a feeble woman!

VITA: Hence you are alive!

SAKARA: Hence you are not killed!

VASANTASENA (*to herself*): Even his courtesy strikes terror! Well, I will proceed thus. (*aloud*) Sir, do you expect to take my ornament?

VITA: Heaven forbid! O Vasantasena, a garden-creeper deserves not to be robbed of its blossoms. So we have nothing to do with ornaments.

VASANTASENA: Then what, indeed, can it be now?

SAKARA: That I, a heavenly personage, a man, and Vasudeva,[32] should be loved.

VASANTASENA (*angrily*): Enough, no more of this! Get away! You talk wickedly.

SAKARA (*clapping his hands and laughing*): Friend, friend! Just see. This courtesan girl is indeed very much attached to me and hence she says to me, 'come, you are fatigued, you are exhausted', while I have not gone to another village, nor to another city. Lady, I swear by my friend's head touched by my feet, I am fatigued and wearied while chasing you here and there.

VITA (*to himself*): Oh! The fool understands *shranta* (fatigued) when she said *shanta* (enough).

(*aloud*) O Vasantasena, you have spoken in a manner contrary to your profession.—

Remember that the courtesans' quarter is dependent on young men for help and also consider yourself like a creeper growing by the wayside. You possess a body that can be bought for money and hence it is like an article for sale. Therefore, O good maiden, serve equally one much coveted by you and one disliked by you.

Moreover—

An erudite eminent Brahmana, and also an idiot of the lowest caste, both bathe in the same well. Even a crow bends on the same

32. Another name for Krishna.

creeper in bloom that had been bent by a peacock sitting upon it. And by the same boat Chandalas (Shudras) cross a river, by which Brahmanas, Kshatriyas and Vaisyas cross it. You are a harlot, and therefore like a well, or a creeper, or a boat. Wait upon all equally.

VASANTASENA: But it is merit indeed that is the cause of love, and not force.

SAKARA: Friend, friend! This accursed wench does not love me, having fallen in love with that penniless Charudatta ever since she visited the garden of Cupid. Here is his house towards the left. So you please manage that she shall not slip from your hands or from mine.

VITA (*to himself*): The fool utters the very thing that should have been avoided! What, Vasantasena has fallen in love with the noble Charudatta? It is indeed well said that 'a jewel unites with a jewel.' Well, then, let her escape, what do I care for this fool? (*aloud*) Son-of-an-unmarried-woman, is that Sarthavaha's[33] house towards the left?

SAKARA: Surely, his house is towards the left.

VASANTASENA (*to herself*): Wonderful! If he is not lying and his house is towards the left, a favour has been conferred upon me by this villain, even when being wicked to me, since it is now possible to meet my beloved.

SAKARA: Friend, friend! In this pitchy darkness Vasantasena has disappeared while being just in sight, like a pellet of lampblack fallen in a heap of black beans.

VITA: Ah! The darkness is indeed very dense. For—long-ranged in light, but suddenly deprived of their power by their entrance into the darkness, my eyes, although wide open have been closed as it were owing to the darkness.

Moreover—

The darkness is, as if, besmearing the limbs; the sky is, as it were, showering down black collyrium[34]; and eyesight has been of no avail, like service rendered unto a wicked man.

SAKARA: Friend, friend! I shall search for Vasantasena.

VITA: Bastard, is there any sign of her that you mark?

33. Refers to a rich man.

34. (Archaic) medicated eyewash.

SAKARA: Friend, friend, of what kind?

VITA: Like the sound of her ornaments or the scent of her chaplets, with its fragrance closely following its track?

SAKARA: I hear the scent of her chaplets: but, my nose being choked up with darkness, I do not very clearly see the sound of her ornaments.

VITA (*aside*): O Vasantasena,—

True that owing to this nocturnal darkness you cannot be seen, like lightning lying concealed in the gap of the interior of a cloud. But, O timid one, this scent proceeding from your chaplets, as well as your jingling anklets, will betray you.

Did you hear me, Vasantasena?

VASANTASENA (*to herself*): I have heard, and comprehended too. (*she shows that she takes off her anklets, and removes her chaplets; then walking about a little, and feeling with her hand*) Ah! Here, surely, is a side-door which I have discovered by feeling the wall. And I know, as it is being closed, that this side-entrance of the house is fastened from inside.

CHARUDATTA: Friend, I have finished my prayers. So you may now go and offer the oblation to the Holy Mothers.

VIDUSHAKA: Sir, I won't go.

CHARUDATTA: Ah! Fie![35]—

Owing to poverty a man's relatives do not abide by his order; his bosom friends turn away from him; his difficulties multiply; his spirit languishes; the brilliance of the moon of his character grows entirely dim; and whatever evil deed may have been done even by other persons, the same is attributed to him.

Moreover—

Nobody keeps company with him, or talks respectfully to him; if he goes to the mansions of the rich on festive occasions, he is looked at, with contempt. Scantily clad, in shame, he keeps himself at a distance from the great. I consider poverty to be, forsooth, the sixth great sin[36].

35. (Archaic) exclamation; used to express disgust or outrage.

36. Referring to the five sins as per Hinduism.

Moreover—

O Poverty, I am sorry just for you, I feel concerned as to where you can go, when I, the unlucky one, shall have departed this body, you having thus so long resided in my frame like a friend.

VIDUSHAKA (*with shame*): Well, friend, if I have to go, then let this Radanika, too, be my companion.

CHARUDATTA: Radanika, follow Maitreya.

MAID: As Your Honour commands.

VIDUSHAKA: Girl, Radanika, take the oblation as well as the lamp. I shall unfasten the side door. (*does so*)

VASANTASENA: The side-door has opened, as though to favour me. Well, then, I shall get in. (*seeing*) Oh! Pity! How now! Here is a lamp!

(*She extinguishes it with the hem of her garment and then enters*)

CHARUDATTA: Maitreya, what is that?

VIDUSHAKA: The lamp has been blown out by the wind coming in with a burst, the side-door being opened. Girl, Radanika, you may go out through the side-door. I too shall follow you after having lighted the lamp from the *antahpura*. (*exit*)

SAKARA: Friend, friend! I shall look for Vasantasena.

VITA: Do look, please, do.

SAKARA (*does accordingly*): Friend, friend! I have caught her.

VITA: Blockhead, it is me!

SAKARA: You should come this side and stand in a corner. (*again searches, and catches* Cheta). Friend, friend! I have caught her!

CHETA: Master, I am Cheta.

SAKARA: Here is Vita, here is Cheta. Vita and Cheta! Cheta and Vita! As for you, you both stand in a corner. (*again searches, and catches* Radanika *by the hair*). Friend, friend, I have really caught her this time. I have caught Vasantasena!—

Like Chanakya seizing Draupadi, I have caught her by her massive hair, as she was indicated by the scent of her chaplets while running away in the dark.

VITA:—

Here, you, who in your pride of youth were going to seek a noble-born person, you have now been seized by the hair, which is richly decorated with flowers and which deserves to be honoured.

SAKARA:—

Here you are caught by your head, O maiden, caught by your hair, by your tresses, by your looks! Now cry or scream or shout very loudly for Sambhu, Siva, Samkara, or Isvara!

RADANIKA (*with alarm*): What are the noble ones bent upon doing?

VITA: Bastard, this consonance of the voice seems to be quite different!

SAKARA: Friend, friend! That whore's-daughter has changed her voice, just as a she-cat changes hers when coveting the cream of curds.

VITA: How! She has changed her voice! Oh! Strange! Or rather, what is so strange about it?—

She has attained skill in her voice, by her acting on the stage, by the training she has received in the arts, and her cleverness in the art of deception.

(*Entering*)

VIDUSHAKA: Ah! Wonderful! Owing to the gentle night-breeze, the lamp is fluttering like the heart of a goat brought in for being tied to the sacrificial post as a victim, (*approaching and seeing* Radanika *in that state*) O Radanika!

SAKARA: Friend, friend! A man, a man!

VIDUSHAKA: Is it indeed proper, is it at all becoming, that the poverty of our noble Charudatta should now make even strangers invade his house with impunity?

RADANIKA: Sir Maitreya, see how I am being insulted!

VIDUSHAKA: What! Is that an insult to you or to us?

RADANIKA: Well, to you, surely!

VIDUSHAKA: Is this violence?

RADANIKA: What else?

VIDUSHAKA: Really?

RADANIKA: Really.

VIDUSHAKA (*angrily raising his wooden stick*): This cannot be allowed! Look here, even a dog becomes formidable when in his own kennel; what then of a Brahmana like me? So, by means of this wooden stick which is crooked like the luck of people like us, I will batter with blows your head, like the top of a spoiled dry bamboo!

VITA: Pardon, worthy Brahmana, pardon!

VIDUSHAKA (*looking at* Vita): He is not the guilty party in this matter. (*looking at* Sakara) *This* fellow, obviously, is the offender in this affair. O you, king's brother-in-law, Samsthanaka[37], bad man, villain! Is this indeed becoming? Even though it is true that the noble Charudatta has been reduced to poverty, still, are not his merits an ornament to this Ujjayini, that you should enter his house and thus insult his servants?—

Insult no man because he is poor, for indeed fate treats none as poor; moreover, even a rich person is poor if he be destitute of good character.

VITA (*feeling ashamed*): Pardon, worthy Brahmana, pardon! This was done, not in insolence, but owing to our having mistaken her for another person. See—

An amorous woman is sought by us—

VIDUSHAKA: What, *this* one?

VITA: Heaven forbid!—

Who is a courtesan. She disappeared, and mistaking this lady for her, we came to commit this breach of decorum.

Please accept this all-in-all of conciliation.

(*he throws away his sword, and folding his hands prostrates himself at his feet*)

VIDUSHAKA: Good man, get up, get up. I blamed you as I did not know your real nature. Now that I know it, I will apologize to you.

VITA: It is indeed you to whom we owe an apology in this affair. So, I will get up, but on one condition.

VIDUSHAKA: Speak up.

VITA: Only if you will not tell Charudatta of this incident.

37. Another name for Sakara.

VIDUSHAKA: I will not tell.

VITA:—

This favour of you, I bear on my head, O Brahmana, we even though armed with real weapons, have been conquered with the weapons of virtues.

SAKARA (*with jealous anger*): Well, for what reason did you fall at this rogue's feet, folding your hands cringingly?

VITA: Because I am afraid.

SAKARA: Of what are you afraid?

VITA: Of the virtues of Charudatta.

SAKARA: What sort of virtues does he possess, when on entering his house one does not find even eatables there?

VITA: Nay, say not so.—He has become impoverished by granting the prayers of persons like us; never did he insult anyone in his pride of wealth. He has dried up by having satisfied the quench of men, like a reservoir, full of water, in the hot season.

SAKARA (*indignantly*): Who is he, that son-of-a-born-slave?—

Is he Svetaketu, the brave and valiant son of Pandu? Or is he Ravana, the Indra-given son of Radha? Or is he Asvatthama, born of Kunti from that Rama? Or is he Jatayu, the son of Dharma?

VITA: You fool! He is, as is well-known, the worthy Charudatta:—

To the distressed, he is the wish-yielding tree, bent down with the load of the fruit of his virtues. To the virtuous he is a family-head. To the learned he is a mirror reflecting learning. He is the touchstone of moral conduct, and the ocean having righteousness for its coast-line. He is ever-hospitable, and never shows disrespect. He is a treasure of all manly virtues, and is courteous and magnanimous by nature. By reason of his manifold virtues, he alone, deserving of praise, really lives; while others are merely breathing as it were.

So we shall go away from this place.

SAKARA: Without taking Vasantasena?

VITA: Lost is your Vasantasena!

SAKARA: How, indeed?

VITA:—

Meeting with you, she has vanished, like the sight of a blind man, or the robustness of a sick person, or the common sense of a fool, or the achievement of an idle fellow, or the highest kind of learning in the ease of one of weak memory and addicted to vices, or like the feeling of love towards an enemy.

SAKARA: I will not go without taking Vasantasena with me.

VITA: Have you not heard even this?:—

'An elephant can be held by means of a tying-rope, a horse by means of reins, and a woman by the heart: if you haven't this, then go away.'

SAKARA: If you want to go, you may. I won't go.

VITA: All right. I am off. (*exit*).

SAKARA: Well, he has taken himself off! (*to* Vidushaka) You villainous fellow, with your head of the shape of a crow's foot! Sit down, sit down!

VIDUSHAKA: We have been already made to sit down.

SAKARA: By whom?

VIDUSHAKA: By fate.

SAKARA: Well, get up, get up!

VIDUSHAKA: We shall get up.

SAKARA: When?

VIDUSHAKA: When fate will again become favourable.

SAKARA: Eh, weep, weep!

VIDUSHAKA: We have been already made to weep.

SAKARA: By whom?

VIDUSHAKA: By poverty.

SAKARA: Eh, laugh, laugh!

VIDUSHAKA: We shall laugh.

SAKARA: When?

VIDUSHAKA: When the noble Charudatta attains prosperity again.

SAKARA: You villainous fellow! Say to that beggar Charudatta at my bidding: 'This courtesan, named Vaṣantasena, decked and covered with gold, and charming like the chief actress rising up for the exhibition of a new play, who has fallen in love with you since her visit to the garden of Cupid, has entered your house while we were trying to win her over by force. So, if you, of your own accord, send her away and place her in my hands, then there will subsist permanent friendliness between you and me, by your quickly restoring her without any judicial proceedings in a court of law and if you don't restore her, there will be a feud lasting till death. Moreover, observe carefully—

A pumpkin whose stalk is besmeared with cow's dung, a vegetable that has been dried, meat which is fried, and boiled rice that has been cooked at night in the cold season—these do not get spoiled even by lapse of time. And so will be my enmity ever-lasting.'

Speak well, and speak craftily. Speak in such a voice that I may hear you when seated in the elephant-corniced upper terrace of my palace. If you speak in any other manner, I will crunch your head like a round *Kapittha*[38] fruit crushed by being inserted between the panels of a door.

VIDUSHAKA: I will speak as you desire.

SAKARA (*aside*): Cheta, has Vita really gone?

CHETA: He has.

SAKARA: Then let us be off quickly.

CHETA: Then let the master take the sword.

SAKARA: Let it be in your own hand.

CHETA: It belongs to the master. Let the master keep it with him.

SAKARA (*seizing it by the wrong end*):—

Like a jackal pursued with yells by dogs and bitches, I go to my house, bearing on my shoulder this unsheathed sword resembling in colour the radish-skin, after making it rest in its scabbard.

(*They walk about and depart*)

VIDUSHAKA: Girl, Radanika, you should not inform Charudatta about this insult offered to you, for I think he would be doubly pained to hear it, afflicted as he already is by poverty.

RADANIKA: Sir Maitreya, I am, indeed, Radanika. My lips are

38. Wood apple or Elephant apple.

sealed.

VIDUSHAKA: Well, so be it.

CHARUDATTA (*to* Vasantasena): Radanika, the boy Rohasena covetous of enjoying the breeze, is now suffering from the chill of the night. So let him be carried inside. And cover him with this cloak. (*he gives her the cloak*)

VASANTASENA (*to herself*): How now! He takes me for his servant! (*taking up the cloak she smells it, then longingly to herself*) Ah! This cloak bears the scent of jasmine flowers! His youth seems to be not indifferent to the pleasures of life, after all. (*she turns aside and covers herself with it*)

CHARUDATTA: I say, Radanika, take Rohasena and go inside.

VASANTASENA (*to herself*): Alas! I have not the good fortune to enter inside.

CHARUDATTA: I say, Radanika, you don't even make a reply! Alas!—

When a man attains a state brought on by fate and painful on account of the loss of prosperity, even his friends turn into his enemies, and even persons who have long been attached to him grow cold.

(Radanika *and* Vidushaka *approach*)

VIDUSHAKA: Sir, *here* is that Radanika.

CHARUDATTA: This is our Radanika! Then who is this another lady, who is desecrated by my garment which I threw at her in ignorance?

VASANTASENA (*to herself*): Say rather *consecrated.*

CHARUDATTA:—

And who appears like the lunar digit obscured by an autumnal cloud?

Or rather, it is improper to look at another's wife.

VIDUSHAKA: Sir, you need not have any fear about looking at another's wife. This is Vasantasena, who has fallen in love with you since her visit to the garden of Cupid.

CHARUDATTA: Ah, this is Vasantasena! (*to himself*)—

Now that my ample fortune has declined, the passion inspired in me subsides in my limbs like the anger of a cowardly person.

VIDUSHAKA: My friend, this is what the king's brother-in-law says—

CHARUDATTA: What is it?

VIDUSHAKA: 'This courtesan, named Vasantasena, decked and covered with gold, and charming like the chief actress rising up for the exhibition of a new play, who has fallen in love with you since her visit to the garden of Cupid, has entered your house while we were trying to win her over by force.'

VASANTASENA (*to herself*): To tell the truth, I have been obliged by these words, 'trying to win her over by force.'

VIDUSHAKA: 'So, if you, of your own accord, send her away and place her in my hands, then there will subsist permanent friendliness between you and me, by your quickly restoring her without any judicial proceedings in a court of law. Otherwise there will be a feud lasting 'till death.'

CHARUDATTA (*contemptuously*): He is a fool! (*to himself*) Ah! How now! She is indeed a lady fit to be waited upon and worshipped like a goddess! Hence it was that at that time,—

She, although ordered to enter the house, did not move away, observing my penurious condition brought upon by fate. Although she can talk boldly in a variety of ways owing to her familiarity with men, she does not speak boldly.

(*aloud*) Madam Vasantasena, I have offended you by thus treating you like a servant, not recognizing you through my ignorance. So here I ask your pardon with my bent head.

VASANTASENA: I have to conciliate Your Honour by bowing down to you, since I have offended you by this intrusion into a place of which I am unworthy.

VIDUSHAKA: Well, both of you happily saluting each other, have mutually approached with your heads, like two paddy-fields. I too will ask pardon of you both, with this head of mine resembling a camel-colt's knee.[39] (*he gets up*)

39. Vidushaka is making a humorous remark here, that like a camel-colt, he is disturbing Charudatta and Vasantasena, who are referred to as paddy fields.

CHARUDATTA: Let it go, no ceremony is needed!

VASANTASENA (*to herself*): The beginning of the talk has been clever as well as sweet. But it will not be proper for me to stay with him, having thus come here, he being in such poor condition. Well, I will speak to him thus. (*aloud*) Sir, if Your Honour would like thus to favour me, then I wish to keep this ornament in your house as a deposit. It is for the sake of ornaments that these wicked fellows pursue one.

CHARUDATTA: This house is not a fit place for keeping a deposit.

VASANTASENA: Sir, that is not true, for deposits are entrusted to *persons*, and not to *houses*.

CHARUDATTA: Maitreya, take that ornament.

VASANTASENA: I am indeed obliged. (*hands over the ornament*)

VIDUSHAKA (*receiving it*): May you be happy!

CHARUDATTA: You fool, it is a deposit and not a gift!

VIDUSHAKA (*aside*): If so, then may thieves steal—

CHARUDATTA: In a very short time—

VIDUSHAKA: This deposit entrusted to us by her—

CHARUDATTA: I will return it.

VASANTASENA: Sir, I wish to return to my house in the company of this gentleman.

CHARUDATTA: Maitreya, accompany Her Ladyship.

VIDUSHAKA: You yourself will look charming, like a royal swan, tripping after this lady with her swan-like gait. But I, a poor Brahmana, will come to grief or perish, being harassed at every step by people, just as an offering made in a public square is gobbled up by dogs.

CHARUDATTA: All right. I myself will accompany Her Ladyship. So light some torches, such as can be relied upon to light up the king's high-road.

VIDUSHAKA: Vardhamanaka, light some torches.

SERVANT (*aside to* Vidushaka): Here, how are torches to be lighted without oil?

VIDUSHAKA (*aside to* Charudatta): Ah! Those torches of ours have now become void of *sneha*[40], like courtesans who, slighting penniless lovers, lose their *sneha*[41] for them.

CHARUDATTA: Maitreya, let it be. There is no need of the torches now. See—

Here indeed, rises the moon, palish-white like the cheek of an amorous lady, having the starry clusters as his retinue, and serving as the lamp on the king's road; the moon, whose white beams fall in the midst of the dense darkness like streams of milk over dried mud from which all the water has oozed out.

(*with love*) Lady Vasantasena, here is your house; you may now get in.

(Vasantasena *looking at him lovingly; exit*)

CHARUDATTA: Friend, Vasantasena has gone in. Come then; we too shall go home.—

Deserted is this royal road, and guards are patrolling it. We must avoid being duped, for night is full of many crimes.

(*walking about*) And this golden casket should be guarded by you at night and by Vardhamanaka by day.

VIDUSHAKA: As Your Honour commands.

(*Both Exit*)

End of Act I, entitled

'The Depositing of the Ornament'[42]

40. Here, oil.
41. Here, affection.
42. Sudraka's *Mrichchhakatika* is the only Sanskrit play in which titles are mentioned at the end of each act. For details on this, read Devendra Raj Ankur's essay included in this volume.

ACT II

(*Entering*)

MAID: The mother has sent me with a message to my mistress, so I will just enter and approach her. (*walking about and observing*) Here is seated my lady, brooding over something in her heart. Well, then, I will approach her.

(*Enter love-sick* Vasantasena *seated, and also* Madanika).

VASANTASENA: Friend, what next?

MADANIKA: Lady, you have not spoken about anything. How then do you ask 'what next'?

VASANTASENA: What did I say?

MADANIKA: You asked, 'what next'?

VASANTASENA (*knitting her eyebrows*): Ah! Yes!

MAID (*approaching*): Lady, the mother asks that you should take your bath and go through the worship of the deities.

VASANTASENA: Friend, tell mother that I am not going to bathe today; let therefore the Brahmana himself perform the worship.

MAID: As Your Ladyship commands. (*exit*)

MADANIKA: Madam, it is my love that prompts me to inquire, and not the disposition to find faults; so tell me, please, what all this means.

VASANTASENA: Madanika, what do you see me to be like?

MADANIKA: I can see from Your Ladyship's absent-mindedness that you have fallen in love with somebody whom you have treasured in your heart.

VASANTASENA: Rightly you have guessed! Clever in fathoming the hearts of others, you are indeed Madanika.

MADANIKA: Well, I am glad, I am glad indeed! Herein you are indeed obliging the mighty Cupid, who is a glorious festive joy to all young persons. So Your Ladyship will please tell whether he is a king, or a king's favourite, who is thus desired to be served!

VASANTASENA: Girl, I want to *enjoy,* not to *serve.*

MADANIKA: Do you love some Brahmana youth, distinguished

for his deep erudition?

VASANTASENA: I have veneration for all Brahmanas.

MADANIKA: Do you then love some young merchant, who has amassed a great fortune by visiting many cities?

VASANTASENA: Friend, a merchant causes great pain of separation, by going to other countries, leaving behind his beloved even though her love for him be very deep.

MADANIKA: Madam, he is not a king, nor a king's favourite, nor a Brahmana, nor a merchant. Who may he be then, that my mistress desires?

VASANTASENA: Girl, you visited with me the garden of Cupid?

MADANIKA: I did, Madam.

VASANTASENA: And still you ask me, as if you are a stranger?

MADANIKA: I got it! Is he that same gentleman who favourably received my mistress when she sought his help?

VASANTASENA: Well, what is his name?

MADANIKA: Why, he resides in the merchants' quarter.

VASANTASENA: Girl, I asked you his *name*!

MADANIKA: My lady, he—of auspicious name—is called the noble Charudatta.

VASANTASENA (*joyfully*): Excellent, Madanika, excellent! You got it rightly.

MADANIKA (*to herself*): Let me speak to her now. (*aloud*) My lady, it is heard that he is a poor man.

VASANTASENA: Hence it is, that I love him. For a courtesan who fixes her affections on a poor lover would not incur censure from the world.

MADANIKA: But, madam, do the honey-bees frequent a mango tree which has lost its blossoms?

VASANTASENA: And hence they are known as 'honey-makers.'

MADANIKA: Lady, if you desire him, then why is it that you do not now approach him clandestinely, at once?

VASANTASENA: Girl, if I were to go to him myself clandestinely, then, being unable to make an adequate return, he may probably make it impossible for me to visit him again, and that I wish to prevent.

MADANIKA: Is it for this very reason that you deposited that ornament in his hands?

VASANTASENA: Girl, you have guessed it correctly!

(*Offstage*)

O Gamester! This gambler, bound to us for ten gold-pieces, is making off—making off! Catch him! Oh! Catch him! Stop, stop! I have seen you!

(*Entering with a toss of the curtain*[1], *and in a flurry*)

SAMVAHAKA: Oh! Painful is this state of a gambler!—

Oh! I am kicked by the *gardabhi* die, as if by a *gardabhi* just let loose from her first halter. Ah! I am struck by the *Sakti* die, like Ghatotkacha struck by the *Sakti* discharged by Karna, the king of the Angas![2]—

Seeing the keeper mentally occupied with the score, I suddenly escaped, and have now reached the public road. Whom shall I now approach for refuge?

Well, then while these two, the master and the gamester, are looking for me elsewhere, I shall enter this deserted temple walking backwards and stand like the idol of goddess. (*he makes numerous gestures and then remains in that posture*)

(*Enter* Mathura, *and the* Gamester)

MATHURA: O Bhattaraka[3]! This gambler here, bound to us for ten gold-pieces, is making off—making off! Catch him! Oh! Catch him! Stop, stop! I have seen you!

1. Yavanika in Sanskrit; a dramatic device used in Sanskrit drama for entry and exit. See Glossary.
2. *Gardhabi* and *Sakti* are the technical names of dice (*cowrics*) used in gambling. It shows that Samvahaka's fate was decided by the dice. There is an analogy of *Gardhabi* and *Sakti* here which also means 'she-ass' and 'weapon' respectively. Ghatotkacha, the son of Bheema, one of the Pandavas in Mahabharata, was killed by Karna, the king of Anga kingdom upon persuasion by Duryodhana, using *Sakti* (weapon), which he had reserved for Arjuna.
3. Refers to the winning gambler.

GAMESTER:—

Even if you now go to the nether regions, or seek the protection of Indra himself, nobody, not even Siva, would be able to save you, excepting the keeper alone!

MATHURA:—

O you who cheated a good keeper! Where, Oh, where can you run away, with your limbs quaking with fear, stumbling at every step over even as well as uneven ground, and seriously disgracing both your family and your good name?.

GAMESTER (*examining the footprints*): Here he goes! Here the track is lost!

MATHURA (*observing and reflecting*): Ah! The footprints are in the contrary direction, while we know that this temple has no image. (*pondering*) That rogue of a gambler has entered the temple, walking backward!

GAMESTER: Then let us follow him.

MATHURA: Let us do so.

(*Both show that they enter the temple; seeing, and making signs to each other*)

GAMESTER: How now! We have a wooden idol here!

MATHURA: No, not at all; it is a stone image. (*he rocks it in various ways; then making a sign*) Nevermind! Come, we shall have a game at gambling. (*they play many sorts of games with dice*)

SAMVAHAKA (*variously restraining his longing to play the game himself; to himself*): Ah!—

The rattling sound of dice stirs the heart of a penniless person, just as the sound of drums does that of a king who has lost his kingdom!—

I know that I shall never again play—gambling is like taking a leap from the summit of the Meru Mountain! But still the rattle of dice, sweet as the song of the Cuckoo, captivates the heart!

GAMESTER: The throw is mine, the throw is mine!

MATHURA: No, no! It is mine, it is mine!

SAMVAHAKA (*coming up suddenly from one side*): No, no! It is *my* throw!

GAMESTER: We have got our man!

MATHURA (*seizing him*): O you, defrauder of the money staked! You are caught! Deliver up those ten gold-pieces!

SAMVAHAKA: I will give today.

MATHURA: Give them just now.

SAMVAHAKA: I will give; please have a little patience.

MATHURA: You fellow, give them *now*, I say!

SAMVAHAKA: My head is feeling dizzy! (*he falls down on the ground*)

(Both *belabour him in various ways*)

MATHURA: Here, now! I arrest you in the name of the Gamblers' Association!

SAMVAHAKA (*gets up; sorrowfully*): How now! I have been arrested in the name of the Gamblers' Association! Ah! This is a rule which we gamblers cannot break. But from what source am I to pay it?

MATHURA: Let a pledge be given!

SAMVAHAKA: So I will. (*approaching the* Gamester) I will pay half; you should forgive me for the other half!

GAMESTER: So be it.

SAMVAHAKA (*approaching the* Keeper): I will give surety for half; Your Honour will please excuse me the other half!

MATHURA: Where is the harm? Let it be as you say.

SAMVAHAKA (*aloud*): Sir, half excused by you?

MATHURA: Excused.

SAMVAHAKA (*to the* Gamester): Half also excused by you?

GAMESTER: Excused.

SAMVAHAKA: Then I am off now!

MATHURA: Give those ten gold-pieces! Where are you off to?

SAMVAHAKA: Look here, Sirs, look here! Oh! Just now I offered security for one half, and the other half was excused. Still he demands it from me, who is unable to pay!

MATHURA (*seizing him*): You rogue, I am Mathura, and am clever enough. I won't allow myself to be cheated in this matter. So, you, cheater-of-the-money-staked, pay up all those gold-pieces *now*!

SAMVAHAKA: How can I pay them?

MATHURA: Sell your father and pay.

SAMVAHAKA: Whence have I a father?

MATHURA: Sell your mother and pay.

SAMVAHAKA: Whence have I a mother?

MATHURA: Sell your own self and pay.

SAMVAHAKA: Then do this much favour: take me to the king's high-road.

MATHURA: Come along!

SAMVAHAKA: So be it! (*walks about*) Sirs, buy me off the hands of this Keeper here by paying ten gold-pieces. (*looking up in the empty space*) What say you?—'What can you do?' I can be a menial servant in your household:—How! He has passed by without making any answer! Nevermind; I will speak to this another gentleman. (*he repeats that same sentence again*) How! He too has despised me and gone away! Alas! This is how I, the unlucky one, have fared since the noble Charudatta lost his fortune!

MATHURA: Give it, I say!

SAMVAHAKA: How can I pay? (*he falls down*)

(Mathura *pulls him up*)

SAMVAHAKA: Save me, sirs, save me!

(*Enter* Darduraka)

DARDURAKA: Oh, gambling is, indeed, to a man, a kingdom without a throne!—

Like a king, a gambler cares not for defeat from any one; he always receives and gives away amounts of money; and, being the recipient of an immense income, he is ever waited upon by wealthy persons.

Moreover—

Money is obtained by gambling alone; wife and friends are acquired through gambling alone; gifts are given and things enjoyed by means of gambling alone; and one's all is lost also by gambling alone.

Moreover—

Having lost all my wealth through the *Treta*, then having my body parched up by the (wrong) fall of the *Pavara*, and next having the road shown to me by the *Nardita*, I now go, being finally fallen to the ground by the *Kata*.[4]

(*looking before him*) Here is our former keeper Mathura, coming in this very direction. Well, I can't escape him, so I will veil myself. (*he gesticulates variously and then stands inspecting his cloak*)—

This garment has lost most of its threads; this garment has become ornamented with hundreds of holes! This garment, it is not possible to wear at all; this garment looks all right only when rolled up!

Or rather, what can this poor fellow do to me, who verily,—

Can remain stretched up, so long as the sun stands in the sky, with one leg in the sky and the other on the ground!

MATHURA: Pay, pay!

SAMVAHAKA: How shall I pay?

(Mathura *pulls him*)

DARDURAKA: Ah! What is this going on? (*looking towards the sky*) What do you say?—'This gambler is being maltreated by the Keeper and none is setting him free'? Why, this Darduraka will set him free! (*approaching*) Make way, make way! (*seeing*) Oh! How now! Here is the cunning Mathura, and this too is poor Samvahaka!—

Why should this very tall and delicate fellow be always engaged in gambling?—he who does not remain suspended motionless till the close of the day, with his head hanging down; whose back does not bear scars produced by rasping clods of earth; and this interior of whose thighs is not daily nibbled by the dogs set after him.

Well, I will first appease Mathura. (*approaching*) I salute you, Mathura!

(Mathura *returns his salute*).

DARDURAKA: What is this?

MATHURA: This fellow owes ten gold-pieces.

4. *Treta*, *Pavara*, *Nardita* and *Kata* are the technical names for particular throws of dice in gambling which means Three, Two, Ace and Four respectively.

DARDURAKA: Why, that's a mere trifle!

MATHURA (*pulling out the cloak lying folded up in* Darduraka's *armpit*): Look, sirs, look! This man, clad in such a tattered garment, calls ten gold-pieces 'a mere trifle'!

DARDURAKA: You fool! I can pay off ten gold-pieces by a mere *Kata* throw. What then! He who has money—does he take it on his bosom and display it publicly? You fellow!—

You are a low-caste man! You are doomed to perish; since you are killing a human being possessed of the five bodily organs, merely for the sake of ten gold-pieces.

MATHURA: Sir, ten gold-pieces may be a trifle to you; to me they are a fortune!

DARDURAKA: If so, then hear me. Give ten additional gold-pieces to this very man, and let him gamble with them.

MATHURA: Then what would happen?

DARDURAKA: If he wins, he will pay you.

MATHURA: And if he doesn't win?

DARDURAKA: Then he will not pay!

MATHURA: Now it is no use talking any more with you in this matter. Rogue, you who are talking in this fashion—*you* pay them! Shall I, too, the well-known gambler, Mathura gamble for nothing? I am not afraid of anyone. You rogue, you are a blackguard[5]!

DARDURAKA: Ha! Who is a blackguard?

MATHURA: You, *you* are a blackguard!

DARDURAKA: Your father is a blackguard! (*makes signs to* Samvahaka *to escape*)

MATHURA: You son-of-a-whore! Is it always thus that you have played at dice?

DARDURAKA: Yes, I have thus played at dice.

MATHURA: You Samvahaka! Pay up those ten gold-pieces!

SAMVAHAKA: I'll pay today; I'll pay today itself.

(Mathura *pulls him*)

5. A man false to his character as a gambler. Refers to a person who behaves in a dishonourable or contemptible manner.

DARDURAKA: You fool! You can harass him in my absence, but not in my presence!

(Mathura *drags* Samvahaka *and gives a blow on his nose;* Samvahaka *shows that he bleeds and faints, and falls on the ground;* Darduraka *goes forth and screens him:* Mathura *strikes* Darduraka; Darduraka *strikes in return*).

MATHURA: Ah! Ah! You villain! Son-of-a-whore! You'll get the fruit of this!

DARDURAKA: You fool, you have struck me while I was on the road; if you were to beat me tomorrow in the palace, then you'll see!

MATHURA: Here I am ready to see!

DARDURAKA: How will you see?

MATHURA (*opening his eyes wide*): Thus I will see.

(Darduraka *throws dust into* Mathura's *eyes and makes a sign to* Samvahaka *to escape;* Mathura *shuts his eyes and falls to the ground;* Samvahaka *makes good his escape*)

DARDURAKA (*to himself*): I have made an enemy of Mathura, the principal gambling-master. So it is not advisable for me to stay any longer in this place. My dear friend Sarvilaka has told me that, as the report goes, a cowherd's son, named Aryaka, shall be the king as has been prophesied by a *Siddha*[6]. And all persons, of my sort, are flocking to him. So I too will go to him. (*exit*)

SAMVAHAKA (*walks about in fear, observing*): Here is somebody's house with its side-door open. So I'll get in here. (*shows that he enters; seeing* Vasantasena) Lady, I beseech your protection!

VASANTASENA: Have no fear, you who seek my protection! Girl, close the side-door.

(Maid *does as ordered*)

VASANTASENA: Whom are you afraid of?

SAMVAHAKA: Of a creditor, madam.

VASANTASENA: Girl, you may now open the side-door.

SAMVAHAKA (*to himself*): How now! She too is scared of the creditors? Rightly it has been said that—

6. A seer

A person, who, knowing his own strength, takes up a load which is proportionate, will never stumble; nor will he perish even if he comes to be in a wilderness.

Of this I myself am an illustration!

MATHURA (*rubbing his eyes*; *to the* Gambler): Eh! Give, give!

GAMESTER: Master, that fellow escaped just as we came into scrape with Darduraka!

MATHURA: That gambler had his nose broken by the blow of my fist. So come along; we shall follow the track of its blood.

(*following the track*)

GAMESTER: Master, he has gone into Vasantasena's house!

MATHURA: Then our gold-pieces are secure.

ANOTHER GAMBLER: Let us go and inform the police.

MATHURA: That rascal might leave this place and go elsewhere. So we better capture him by means of a blockade.

(Vasantasena *signals to* Madanika)

MADANIKA: Whence do you come? Who may you be? To whom do you belong? What profession do you follow for a living? And whence is your fear?

SAMVAHAKA: May the lady hear!—Madam, my birthplace is Pataliputra, and I am the son of a village headman. I earn my living by the profession of a shampooer.

VASANTASENA: It is a very dainty art that you have learnt.

SAMVAHAKA: Lady, I learnt it as an art, but it has now become my means of earning a livelihood!

MAID: You have given a reply indicative of extreme dejection! Well, what next?

SAMVAHAKA: Then, lady, having at home heard about this city from the lips of travelers, I came down to this place out of a curiosity to visit a country never seen before. And, having entered this Ujjayini, I took service with a certain gentleman. He is so very handsome in appearance and of a sweet address! He makes gifts, but makes no mention of them; and forgets any ill. Why waste words?—On account of his courtesy he considers himself as belonging to *others*; and he is kind to those who approach him for refuge.

MAID: Who may he be that now graces Ujjayini, having appropriated to him the virtues of the object of my lady's desires?

VASANTASENA: Well spoken, wench, well spoken! I too thought just the same in my mind.

MAID: Sir, what next?

SAMVAHAKA: Lady, owing to the large gifts he bestowed in compassion, he now—

VASANTASENA: Has lost all his fortune?

SAMVAHAKA: How did Your Ladyship know it, even before it was told to you?

VASANTASENA: What is there to be known? Virtues and riches are seldom found to exist together. Water is most plentiful in those pools which are unfit to be drunk from.

MAID: Sir, what may his name be?

SAMVAHAKA: Lady, who is there that does not know the name of that Moon on this Earth? He resides in the merchants' quarter; he, of auspicious name, is called the noble Charudatta.

VASANTASENA (*joyfully getting down from her seat*): This house is Your Honour's own! Girl, give him a seat. And take up a fan; His Honour is weary with fatigue.

(Maid *does as ordered*)

SAMVAHAKA (*to himself*): How! By the mere mention of the noble Charudatta's name they are showing me so much respect! Excellent, noble Charudatta, excellent! You alone really *live* in this world, while all other persons merely *breathe*! (*falling* at Vasantasena's *feet*) No, no, madam! You may sit on the seat!

VASANTASENA (*sitting down on it*): Sir, whence is that Charudatta be a rich man?

SAMVAHAKA:—

A good man's wealth really consists in his hospitable nature; otherwise who does not possess riches which are transitory? He who knows how to respect others knows also the significance of the respect paid by others to him.

VASANTASENA: Well, go on.

SAMVAHAKA: Then that gentleman took me into his personal service on wages; when he had only his good name left to him, I took to gambling as a means of sustenance. Then, my luck going against me, I lost ten gold-pieces at play.

MATHURA (*from a distance*): I am totally undone! I am robbed!

SAMVAHAKA: Here are they, the keeper and the gamester, both on my track. So now, having heard my story, Your Ladyship may decide what to do.

VASANTASENA: Madanika, the birds wander anywhere at random, when the tree whereon they dwell is tottering. Go, then, girl, and give this bracelet to those two, the keeper and the player, saying that it is given by this gentleman himself. (*she removes the bracelet from her hand and gives it to the* Maid)

MAID (*taking it*): As Your Ladyship commands. (*exit; goes to* Mathura)

MATHURA: I am totally undone! I am robbed!

MAID: As these two are looking upwards, sighing deeply, pondering, and talking with their eyes fixed on our door, I guess that these must be those very persons, the keeper and the gamester. (*approaching*) Sir, I bow to you!

MADANIKA: May you fare well!

MAID: Sir, which of you two is the Keeper?

MATHURA: O you slender-waisted damsel! With your lower lip bitten in love sport and looking ungainly, to whom are you addressing alluring words, casting side-glances?

I am not a rich man. So go elsewhere.

MAID: If you can talk thus, then surely you can't be the gambler. —Is there any person who owes you money?

MATHURA: There is. He owes ten gold-pieces. What of him?

MAID: On his account my lady has given this bracelet—Ah! No! —he himself has given it.

MATHURA (*joyfully taking it*): Aye, you say to that good chap that we have got the security; so he might come and again have a throw at dice.

(*Exit both*)

MAID (*approaching* Vasantasena): Mistress, the keeper and the gamester have both gone away satisfied.

VASANTASENA: Then you may depart, sir, and comfort your anxious relatives.

SAMVAHAKA: Madam, if so, then kindly allow this art of mine to be imparted to your maid-servants.

VASANTASENA: Sir, you should go and serve that same person whom you served before, for whose sake this art is to be learnt.

SAMVAHAKA (*to himself*): Her Ladyship has skillfully declined my offer! How then shall I repay her kindness? (*aloud*) Madam, being thus disgraced as a gambler I will now become a Buddhist monk. So madam should carefully remember these words.

VASANTASENA: Sir, don't be rash.

SAMVAHAKA: Madam, my resolve is taken. (*walking about*)—

Gambling has done to me that which places me beyond the power of all people. Now I will roam about on the king's high-road freely holding up my head.

(*An uproar offstage*)

SAMVAHAKA (*listening*): Ah! What may this be? (*looking towards the sky*) What say you?—'Here has broken loose Vasantasena's rogue elephant named Khuntamodaka?' Oh! I shall then go and have a look at Her Ladyship's Scent-elephant.[7] Or rather, what have I to do with this? I shall do as I have resolved. (*exit*)

(*enter* Karnapuraka *with a toss of the curtain; highly jubilant and in a formidable and gaudy dress*)

KARNAPURAKA: Where, where is Her Ladyship?

MAID: You rude fellow, what is the cause of your excitement that you do not see Her Ladyship seated just in your front?

KARNAPURAKA (*seeing*): Madam, I bow to you!

VASANTASENA: Karnapuraka, you seem to be mightily pleased,

7. One of the species of elephants with a strong scent which repels ordinary elephants.

from your face! So what can this be?

KARNAPURAKA (*with great pride*): My lady, you've missed a treat inasmuch as you saw not Karnapuraka's exploit, today!

VASANTASENA: Karnapuraka, what is it, what is it?

KARNAPURAKA: May Your Ladyship listen—that rogue elephant of yours, called Khuntamodaka, he broke the post to which he was tied, killed his keeper, and got down to the king's high-road, causing tremendous commotion. Then in the meantime people began to shout out—

"Remove the children—Quickly mount trees or house-tops: Don't you see that the furious elephant is just coming towards you this way?"

And again—

"Pairs of anklets have dropped off! Broken are girdle-zones set with jewels! And so, too, are very lovely bracelets inlaid with a network of the shooting rays of gems!"

Then that rogue elephant came upon a monk, as he was dashing through the city of Ujjayini, as though it were a lake full of blooming lotuses through which he was wading with his trunk, legs and tusks. Seeing that he discharged a shower of spray at the monk who had already, in fear dropped his staff, his gourd, and his alms-pot, and took him up between his tusks, the people again shouted out—"Oh! The monk is being killed!"

VASANTASENA (*in agitation*): Oh! The carelessness! Oh! The mishap!

KARNAPURAKA: Away with agitation! May Your Ladyship just listen: Then seeing the elephant who carried along a mass of broken chains that were dangling about and bore that monk held between his tusks, I, Karnapuraka—no, no! This slave nourished upon the rice-balls got from Your Ladyship—running haphazardly, calling out to that writer of gambling-scores, quickly seized an iron bar from a shop and challenged that mad elephant.

VASANTASENA: What happened then?

KARNAPURAKA:—

I struck at that infuriated elephant who looked like a peak of the Vindhya mountain, and released that recluse held between his tusks.

VASANTASENA: It was bravely done by you! What followed?

KARNAPURAKA: Then, lady, like a boat unevenly loaded with freight, all Ujjayini crowded over on one side, simply calling out 'Bravo! Karnapuraka, Bravo!' Then, lady, one person from amongst them, touching these parts of his body where he should have had ornaments, and finding that they weren't there, looked up, heaved a deep sigh, and threw this cloak over me.

VASANTASENA: Karnapuraka, just see whether that cloak is scented with jasmine flowers or not.

KARNAPURAKA: Lady, owing to the smell of ichor, I cannot well distinguish its scent.

VASANTASENA: Just see if it has any name.

KARNAPURAKA: Here's a name. May madam read it herself. (*takes the cloak to her*)

VASANTASENA: 'This cloak belongs to the noble Charudatta.' (*reading it she takes it lovingly and wraps herself with it*)

MAID: Karnapuraka, the cloak becomes our lady well.

KARNAPURAKA: Surely; it becomes her well!

VASANTASENA: Karnapuraka, here's a reward for you. (*gives him an ornament*)

KARNAPURAKA (*receiving it with bowed head, and saluting her*): Now that cloak does, indeed, become Her Ladyship beautifully!

VASANTASENA: Karnapuraka, where may the noble Charudatta be at this time?

KARNAPURAKA: He started to go home by this very road.

VASANTASENA: Wench, we shall ascend the upper terrace and have a look at the noble Charudatta.

(*Exit all*)

Thus ends Act II, entitled
'Samvahaka, the Gambler'

ACT III

(*Enter* Cheta)

CHETA:—

A master, who is kind and compassionate to his servants, is pleasing even though he may be poor. But again, one who is mean and purse-proud is forsooth difficult to serve, and proves formidable in the end.

Moreover—

It is not possible to turn away a bull who is greedy of corn, nor a man who is deeply attached to another's wife, nor one who is addicted to gambling; nor is it possible to cure what is one's natural defect.

A long time has passed since the noble Charudatta went to attend the musical concert. It is past midnight, and still he hasn't come. So then I'll go to the room near the outer entrance, and go to sleep there. (*he does accordingly*)

(*Enter* Charudatta *and* Vidushaka)

CHARUDATTA: Oh! Splendid! Rebhila sang just beautifully! The lute is indeed a jewel, only that it was not produced from the ocean. For—

To a love-sick person, it is a companion agreeable to the state of his heart; if the lover who has made an appointment is long in coming, it serves as an excellent means to pass time; to those suffering from the pangs of separation, it is the most welcome means of consolation; and to a person actually in love, it yields delight that intensifies his passion.

VIDUSHAKA: Sir, come along; let us go home.

CHARUDATTA: Master Rebhila did, indeed, sing charmingly!

VIDUSHAKA: As for me, a smile is produced in me by two things: by a woman reading Sanskrit, and by a man singing in a low, sweet tone. For when a woman reads Sanskrit, she noisily snuffles, like a young cow that has had a new rope put through the septum of her nose. A man, too, when he is singing in a low, sweet tone, is emphatically not to my liking, like an old priest repeating *mantras* and having a garland of dried-up flowers round his neck.

CHARUDATTA: Friend, Master Rebhila did indeed, sing wonderfully today! And how is it that you were not pleased?—

Really his song was impassioned, sweet, smooth, clear, and full of emotion, and graceful, and charming. Or where is the use of many words of praise uttered by me? Methinks it was as if some woman concealed who sang, and not he.

Moreover—

To tell the truth, I feel that, although the time of the musical concert is over, I am as it were still listening, as I walk, to the cadence of the notes of his sweet voice; and to those well-blended notes of the lute, high-pitched when following up with the harmonious swell and fall of the sounds of the syllables of the song, but low towards the close; and to the song, sung with due regard to the pitch of sounds, graceful and repeated twice where the *raga*[1] required it.

VIDUSHAKA: Friend, even the dogs are fast asleep in the various lanes that stretch across the market-place. So let us go home. (*looking before him*) Friend, see, see! Here is the divine Moon descending from his skyey[2] mansion, as though to make room for darkness!

CHARUDATTA: Your remark is quite proper.—

There sets the moon, giving room to darkness, and with one end raised upwards, looking as if it were the sharp tip remaining visible outside, of the tusk of a wild elephant submerged in water.

VIDUSHAKA: Well, here's our house. Vardhamanaka, Vardhamanaka! Open the door!

CHETA: I hear the revered Maitreya's voice; the noble Charudatta has arrived. Well, then, let me open the door for him. (*he does so*) Sir, my bow to you! Maitreya, to you too! May you both sit on this commodious seat!

(*Both enter and sit down*)

VIDUSHAKA: Vardhamanaka, call Radanika to wash our feet.

CHARUDATTA (*with compassion*): You need not awaken a person who is sleeping.

1. Repeated through enthusiastic fondness or passion for music.
2. Like the sky; ethereal, being in the sky.

CHETA: Sir Maitreya, I'll hold the water, and you may wash Charudatta's feet.

VIDUSHAKA (*angrily*): Friend, here is this fellow, the son-of-a-slave, going to hold the water, while he wants to make *me*, who am a Brahmana, wash your feet!

CHARUDATTA: Friend Maitreya, you may hold the water and Vardhamanaka will wash my feet.

CHETA: Sir Maitreya, pour the water.

(Vidushaka *does accordingly,* Cheta *washes* Charudatta's *feet and moves away*)

CHARUDATTA: You should also pour water over the Brahmana's feet.

VIDUSHAKA: What good will water for my feet do? I shall again soon have to roll on the ground, like a beaten ass?

CHETA: Sir Maitreya, you are indeed, a Brahmana!

VIDUSHAKA: Yes, I am a Brahmana among all the Brahmanas, like the harmless and despised *dundubha*[3] among all the serpents!

CHETA: Sir Maitreya, still I shall wash your feet. (*does so*) Sir Maitreya, here's that golden casket which I have to guard by day and you by night. So please take it. (*gives it; exit*)

VIDUSHAKA (*taking it*): This still remains with us! What, is there not even a thief in this Ujjayini who would rid me of this vile thing, this robber of my sleep? Friend, I am going to keep it inside, in the *antahpura*.

CHARUDATTA:—

Do not take it inside the quadrangle, for it has been worn by a courtesan; therefore, O Brahmana, you should keep it with you till it is returned to her.

(*Showing that he is about to sleep, he again recites the former verse, 'To tell the truth' et al*)

VIDUSHAKA: Are you going to sleep?

3. Vidushaka unmindfully compares himself to a *dundubha* for which nobody cares—implying that being a Brahmana without the qualifications of one, he deserves no regard from anyone.

CHARUDATTA: Yes.—

Here is this sleep approaching me as it were from my forehead, seizing my eyes; of invisible form and elusive, it gains strength, like old age, overcoming the opposing vitality of a person.

VIDUSHAKA: Well, we shall both sleep (*shows that he falls asleep*).

(*Enter* Sarvilaka)

SARVILAKA:—

Having made a passage for my business, by the power of my training and physical strength—a passage large enough to easily admit a man of my build—I move about, my sides being scratched by my creeping along the ground, like a snake with his old worn-out body casting off his slough.

(*joyfully looking at the sky*) Ah! How now! The divine Moon is about to set! For now—

This night, in which all the objects are hidden from sight by the pitchy darkness, is concealing me as fondly as a mother conceals her son, a mother in whom all other sentiments are overpowered by the intense blindness of love; I, who walk about in fear of the king's watchmen, and who, supremely brave, am bent upon burgling the houses of others.

Having made a hole in the surrounding wall of the park, I have now reached the central building. So now I shall force an entry into the *antahpura* also. Oh!—

Let people call, as they will, this a vile trade, which thrives we when people are asleep; and that getting the better of people by cheating them when they are unsuspicious, is mere thieving, and by no means a brave deed. Still, even a condemnable position of independence is preferable to serving others with folded hands. And this is the path that was followed of yore by Drona's son[4] when he assassinated the Kashatriya prince in sleep.

So now, in what spot shall I effect a breach?—

What is that spot, which is loose soft by being sprinkled over with water, so that it would produce no sound when being bored through?

4. Asvatthaman; a reference from the *Mahabharata*.

Where could be made a hole in the wall, wide but not in sight of the passers-by? What portion of this building is dilapidated, having its masonry worn out by the corroding action of saline exudations? And where shall I not come upon women-folk, and still accomplish my purpose?

(*touching the wall*) Here's a spot, worn out by saline action, and weakened by being sprinkled every day with water at the sight of the sun. And here's a pile of rubbish excavated by rats! My purpose in hand here is as good as accomplished! For this is the best omen of success for us thieves, the sons of Karttikeya[5], our patron-saint. Now I have to begin my job; what sort of hole shall I make here? In connection with this the divine Kanakasakti[6] has laid down four modes of making a breach they are as follows: to pull out the bricks when they are baked, to chip them off if unbaked, to wet them with water if they are of common clay, and to cut them through if they are of wood. Now this is a wall of baked bricks, so I shall have to pull them out. Here seven kinds of holes are possible;—

Padmavyakosa, Bhaskara, Balachandra, Vapi, Vistirna, Svastika and Purnakumbha.[7] So in what spot shall I exhibit my skill, seeing which tomorrow the citizens would be filled with astonishment?

Well, in this particular case of a baked-brick wall, the Purnakumbha breach alone would look well. I will make that.—

When at night other walls were breached by me, as also hard structures, weakened by the action of saltpetre, the neighbours, seeing them in the morning, have condemned my crime, and praised my skill

5. Science of thieving is ascribed to Yogacharya, who was taught the science by Karttikeya, son of Siva. He is worshipped by people for the recovery of stolen property. It is for this reason perhaps that thieves paid homage to him.
6. The wielder of the golden spear in Hindu mythology.
7. Seven different shapes to make breaches in a brick wall, included in the art of thieving—
Padmavyakosa: Having the shape of a full-bloomed lotus; *Bhaskara*: Like the sun; *Balachandra*: Cresent-shaped; *Vapi*: Round-shaped; *Vistirna*: Oblong-shaped; *Swastika*: Shape of a swastika, an auspicious symbol—cross with ends bent round; *Purnakumbha*: Literally, "full or complete vessel", it is an important part of Hindu rituals. Here, it seems to refer to the shape of such a vessel.

of handiwork as well.

My bow to Kumara Karttikeya, the bestower of boons! Reverence to Kanakasakti, the deserving of praise, whose law is divine! Salutation to Bhaskaranandin! And reverence to Yogacharya,[8] whose first pupil I am. Highly pleased with me, he gave me magic ointment.—

When I am besmeared with it, the watchmen will be unable to see me; nor will any weapon striking my body, produce pain.

(*he does accordingly*) Ah! Alas! I have forgotten my measuring tape! (*reflecting*) Well, this sacred thread will serve as the measuring tape! Indeed, the sacred thread is a very important serviceable article to a Brahmana, particularly to one of my profession! For—

With this, one measures the passage to be made in the wall; with this one breaks off the fastenings of ornaments; this serves as a key when a door is securely locked; and it also serves as a ligature when one is bitten by insects or snakes.

So, I'll take measurements and start on my work. (*does so, then inspecting*) Only one brick remains to be taken out from this breach. Ah! Pity! I am bitten by a cobra! (*he binds up his finger with his sacred thread, and shows that he is affected by the poison, then applying the cure*) I am all right now! (*again works at the breach, observing*) Ah! A lamp is burning!

For—

This flame of the lamp, yellowish like gold, emerging through the hole and lying across the ground, being surrounded by darkness on all sides, looks like a streak of gold scratched on a touchstone.

(*again working at the breach*) Here's the hole finished! Well, let me get in. Or rather, I won't go in at once; I will first shove in a dummy man. (*does so*) Ah! There's nobody about! Bow to Karttikeya! (*enters, observing*) Ah! Here are a couple of men asleep! Well, I'll open the door to provide a means of escape for myself. How! The house being old, the door creaks! Well, now, I'll first look for water. Where indeed might water be? (*searches here and there; finding water, he sprinkles*

8. It is believed that Kanakashakti, Bhaskaranandin and Yogacharya are the three writers on, and the teachers of, the art of thieving. Kanakashakti is the wielder of the golden spear. Bhaskaranandin may mean 'whose law is divine.' Yogacharya is the reputed author of the treatise on the black art.

it about; then with apprehension) I hope it won't make a noise when being sprinkled on the ground. Well, now, I will do this. (*looking behind, he opens the door*) Well, this is all right. Now I'll see if these two are feigning sleep or are really asleep. (*frightening them, and observing*) Ah! They must be really asleep. For—

Their breathing is free, is quite distinct, and occurs at regular intervals. Their eyes are firmly closed; they neither blink nor are they unsteady behind the lids. Their limbs lie loose, owing to the bodily joints being relaxed, and they. stretch beyond the limits of the bed. Moreover, if they were shamming sleep, they could not have borne the gleam of the lamp held before their face.

(*looking about him*) Ah! How now! Here is a tabour; here a flute, and here a *panava*! Here too is a lute;[9] here are reed pipes and here these puppets! How! This turns out to be the house of a professor of music! Or why, I have entered by trusting to the external grandeur of the mansion.—

Now, is he really a poor man, or does he keep his money underground, from fear of the king or of thieves? Ah, can there be a treasure buried to Sarvilaka? Well, I'll scatter the magical seeds.[10] (*does so*) The seeds, when scattered, nowhere multiply or expand. Oh! He must really be a poor man! Well, I'll depart.

VIDUSHAKA (*talking in his sleep*): Friend, I see something like a hole! I see someone like a thief! So take hold of this casket containing gold ornaments.

SARVILAKA: What! Can it be that he, knowing that I entered here, is mocking me that he is poor! Shall I then kill him?—Or is he talking in sleep owing to lightheadedness? (*seeing*) Ha! This thing, wrapped up in a worn-out bathing-garment, is really a casket of gold ornaments, as it gleams in the light of the lamp. All right; I'll take it —Or rather, it is not proper to rob a person of high birth but poor like myself. So I'll go away.

9. Reference to various musical instruments, suggesting Charudatta's love for music.
10. According to a treatise on house-breaking, enchanted (magical) seeds expand or multiply when scattered on the ground where there is treasure buried underneath.

VIDUSHAKA: Friend, I adjure you by the wishes of cows and Brahmanas; do take this gold-casket.

SARVILAKA: The adjuration by the saintly wishes of cows and Brahmanas must not be transgressed. So I'll take it. But no! Here's a lamp burning! I carry with me a moth[11] for extinguishing lamps. So I'll let it loose; this is the proper place and time for it. Here this insect, set free by me, flies straightway and whirls over the lamp, describing wonderful circles. Here is the lamp extinguished by Bhadrapitha with the flutter of its wings! Oh! It has caused darkness! Or rather, fie on me, who too have caused darkness[12] in our Brahmana family! For here am I, a Brahmana, Sarvilaka by name, the son of one who knew by heart all the four Vedas and who never accepted any gifts, doing this foul deed for the sake of the courtesan Madanika! Now I will grant the Brahmana's request. (*proceeds to take the casket*)

VIDUSHAKA: Friend, your fingers are cold!

SARVILAKA: Oh! What carelessness on my part! My fingers are cold owing to their recent contact with water. Well, I'll hold my hand for a while inside my armpit. (*he acts warming his right hand, and then takes the casket*).

VIDUSHAKA: Have you got it?

SARVILAKA: Of course, I couldn't say 'no' to this request of a Brahmana; so I've accepted it.

VIDUSHAKA: Now I can sleep soundly, like a merchant that has sold off his goods.

SARVILAKA: Worthy Brahmana, you may sleep for a hundred years! Alas! For the sake of the courtesan Madanika, I have thus brought degradation upon my Brahmana family; or why, I have degraded my own soul?

Fie upon poverty, on account of which one's manly nature ceases to feel disgust for unworthy acts! For instance, here I am censuring this ignoble deed, but am doing it all the same!

Well now let me return to Vasantasena's mansion to buy Madanika's freedom. (*walking about and observing*) Ah! It sounds like somebody's

11. Name of an insect.

12. Here, in the sense of 'brought ignominy upon'.

footsteps! I hope they are not policemen! Nevermind—I'll become rigid like a pillar and stand here. Or rather—are the police of any account to me Sarvilaka? I, who—am a veritable cat in leaping, a deer in bounding off, a hawk in seizing prey and tearing it to pieces, a dog in judging of the strength of a man according as he is asleep or awake, a snake in crawling, magic personified in assuming different characters, postures and dresses, the goddess of speech in the various dialects of different countries, a lamp during nights, a *dudubha* in slipping away from intricate places, a horse on land, and a boat on water!

Moreover—

In movement I am quick like a snake; in steadiness, like a mountain; in flying, I resemble the lord of birds[13]. In surveying the whole country, I am like a hare; in effecting a capture, like a wolf; and in strength, a lion.

RADANIKA (*entering*): Oh! Alas! How now! Vardhamanaka was sleeping in the room adjoining the outer gate; he, too, is not to be seen here! Well, I'll call out to the worthy Maitreya. (*walks about*)

SARVILAKA (*prepares to strike down Radanika; but observing*): Ah! She is a woman! Nevermind, then, I am going. (*exit*)

RADANIKA (*moving on, with fear*): Ha! What's that! A thief has burgled our house and is escaping! Well, I'll go and rouse Maitreya. (*approaching Vidushaka*) Sir Maitreya, get up, get up; in our house a thief has cut a hole and escaped!

VIDUSHAKA (*getting up*): Ah! What do you say, you slave-born wench? A hole has cut a thief and escaped?

RADANIKA: You silly fellow, don't you be joking; can't you see this hole?

VIDUSHAKA: Ah! What do you say, you slave-born wench? That he has cut open a second door as it were? Here, friend Charudatta, get up, get up; a chief has cut a hole in our house and escaped!

CHARUDATTA: Enough; don't you be joking.

VIDUSHAKA: My dear friend, this is no joke; you should see for yourself.

CHARUDATTA: In what place?

13. The eagle.

VIDUSHAKA: Here it is.

CHARUDATTA (*observing*): Ah! It is a fine looking hole!—

This breach, formed with the bricks pulled out from the upper part downwards, which is narrow at the top and wide in the central part, looks as if it were the heart of our stately mansion, burst in fear at coming in contact with an unworthy person.

How! There is a certain skill even in such kind of work!

VIDUSHAKA: Friend, this hole could have been made only by two sorts of persons; either by a stranger, or by one who wished to try his hand; for otherwise, who is there in this Ujjayini that does not know the wealth of our house!

CHARUDATTA:—

This hole in our house must have been made by a foreigner who wanted to practise his art; he did not know that people here were sleeping soundly because they were moneyless! At first, on seeing the magnificent structure of our house, he was filled with hopes and entered; but he must have afterwards gone disappointed, being just fatigued, having long toiled at boring that hole.

So what will that poor fellow say to his friends?—that he entered the house of a leading merchant's son, Charudatta, but got nothing?

VIDUSHAKA: Sir, how is it that you feel sorry for that rascal thief? He must have thought—here is a big mansion; from here I shall carry off a casket of jewels or of gold—(*remembering sorrowfully; to himself*) Ah! Where's that golden casket?—(*again remembering, aloud*) Friend, you are always saying that Maitreya is a fool, that Maitreya is a blockhead. But I acted wisely in that I delivered over that golden casket into your hands! Otherwise that son-of-a-slave would have taken it away.

CHARUDATTA: Enough of joking.

VIDUSHAKA: Sir, though I may be a fool, still do I not know even the time and place for joking?

CHARUDATTA: At what time did you give it to me?

VIDUSHAKA: Sir, just when I said to you that your fingers were cold.

CHARUDATTA: Perhaps it might have been even as you say—

(*looking everywhere, joyfully*) Friend, good luck! I'll tell you something pleasant!

VIDUSHAKA: What, it is not stolen?

CHARUDATTA: It is.

VIDUSHAKA: Then what is it that is pleasant?

CHARUDATTA: That he went away satisfied.

VIDUSHAKA: But it was a deposit!

CHARUDATTA: Ah! A deposit? (*faints*)

VIDUSHAKA: Come to yourself, Sir. If a thief has stolen the deposit, why should you faint?

CHARUDATTA (*recovering*): Friend,—

Who will believe the real fact? Everyone will think lightly of me. For, in this world, inglorious poverty is always prone to be suspected.

Ah! Bad luck!—

If, indeed, Fate did desire my riches, why did now the cruel one spoil my character also?

VIDUSHAKA: I would, indeed deny it and say—Who made the deposit? Who accepted it? Who was the witness?

CHARUDATTA: Am I now to tell a lie? No.—

I'll again acquire the means of returning the deposit, even by begging if necessary; but I'll never utter a falsehood, which would lead to a loss of character.

RADANIKA: Well, then, let me go and tell all this to the worthy Dhuta. (*exit*)

(*Enter* Charudatta's *wife, accompanied by the* Maid)

WIFE (*excitedly*): Well, is my lord really safe in body, along with the worthy Maitreya?

MAID: Mistress, he is really safe. But that ornament belonging to the courtesan, that is stolen.

(Wife *swoons*)

MAID: Worthy Dhuta, have patience!

WIFE (*recovering*): Girl, What did you say?—that my lord was safe in body? It were now far better that he was injured in body, than in character. For now in Ujjayini people will say that my husband himself; being poor, committed such a vile deed. (*looking upwards and sighing*) O mighty Fate! You do, indeed, play with the fortunes of poor people, unstable as the drops of water fallen on lotus-leaves! Now, I have only a jewel-necklace left, come from my mother's family. But even this, my husband would refuse to accept, on account of his extreme pride! Girl, just call the worthy Maitreya here.

MAID: As the noble Dhuta orders. (*approaching* Vidushaka) Sir Maitreya, Dhuta is calling you.

VIDUSHAKA: Where is she?

MAID: Here is she; go to her.

VIDUSHAKA (*approaching*): My blessings to Your Ladyship.

WIFE: Sir, my bow to you! Sir, just turn your face towards the east.

VIDUSHAKA: Lady, here I have turned my face towards the east.

WIFE: Sir, take this.

VIDUSHAKA: What may this be?

WIFE: I had observed a fast on Ratnashashthi day[14]. For that I had to bestow a gift on a Brahmana, in accordance with my position. But no Brahmana was yet given a gift. So you should accept this jewel-necklace from me on his account.

VIDUSHAKA (*receiving*): God bless you! Now I will go and tell this to my dear friend Charudatta.

WIFE: Sir Maitreya, but take care that you do not put me to shame thereby! (*exit*)

VIDUSHAKA (*with wonder*): Oh! What a nobility of mind she shows!

CHARUDATTA: How now! Maitreya yet tarries. I hope he won't do anything rash in his present distracted condition. Maitreya! Maitreya!

VIDUSHAKA (*approaching*): Here I am. Take this. (*shows the jewel-necklace*)

CHARUDATTA: What's this?

VIDUSHAKA: Sir, this is what I would describe as the fruit of your having married a worthy wife!

14. The ceremonious day when a jewel is presented to a Brahmana.

CHARUDATTA: Alas! Even my wife takes pity on me! Ah! Now I am truly a poor man!

I, having lost my money on account of my own bad luck, am now shown pity by my wife's money! Through the absence of money, a man becomes a woman; and she who is a woman becomes a man, by giving the money!

Or rather, I am not a poor man, I, whose—

Wife follows me through the turns of my fortune; who have a friend in you, unchanged in happiness as well as in sorrow, and who have not swerved from truth, which is rarely to be found in the poor.

Maitreya, go with that jewel-necklace to Vasantasena and say to her at my instance: 'Kindly accept this jewel-necklace in exchange for that golden casket of yours, which we thought to be our own and in that belief lost at gambling.'

VIDUSHAKA: Don't you give away this jewel-necklace, which is, as it were, the best produce of the four oceans, for the sake of that thing of comparatively little value, which was neither swallowed nor enjoyed but was stolen by thieves!

CHARUDATTA: Nay, friend, say not so.

This price is being paid, not for that ornament, but for that high trust, by relying on which she kept that deposit with us.

So friend, I adjure you by the touch of my body—you must not return here without inducing her to accept it. Vardhamanaka!

Quickly fill up the opening wall with these bricks; I do not wish it to remain; for scandal has many evils.

Friend Maitreya, you should also address her with due dignity and nobility of mind.

VIDUSHAKA: Friend, can one talk ignobly, even though one may be poor?

CHARUDATTA: Friend, I am not indeed poor, I whose (*repeats again the verse, 'Wife follows me' et al*). So you may go now. I too will purify myself and attend to morning prayers.

(*Exit all*)

Thus ends Act III, entitled

'The Making of a Breach'

ACT IV

(*Enter* Maid)

MAID: The mother has asked me to be in attendance upon Her Ladyship. Here she is, talking about something with Madanika, her eyes fixed on a painting-board. Well, then, I'll approach her. (*walks about*)

(*Enter* Vasantasena *and* Madanika)

VASANTASENA: Girl, Madanika, is this portrait a most faithful representation of the worthy Charudatta?

MADANIKA: It is a most faithful one.

VASANTASENA: How do you know?

MADANIKA: Because your eyes are gazing upon it with great fondness.

VASANTASENA: Madanika, do you speak thus because of the courtesy bred in you by your residence in the courtesans' quarter?

MADANIKA: My lady, is it a fact that every person who lives in the courtesans' quarter is falsely courteous?

VASANTASENA: Girl, courtesans *are* false in their courtesy on account of their coming in contact with various sorts of men.

MADANIKA: Since in the first place Your Ladyship's eyes as well as heart take delight in this portrait, where is the need to ask the reason?

VASANTASENA: Girl, I wish to avoid incurring the ridicule of my friends.

MADANIKA: It couldn't be so, my lady, for we women always act in accordance with the wishes of our friends' hearts.

FIRST MAID (*approaching*): Mistress, the mother says : 'A veiled carriage is ready, waiting at the side-door; so you should go.'

VASANTASENA: Wench, has the noble Charudatta sent for me?

MAID: Mistress, it is the same who has sent an ornament worth ten thousand gold-pieces along with the carriage.

VASANTASENA: But *who* is he?

MAID: The king's brother-in-law, Samsthanaka.

VASANTASENA (*angrily*): Get away! Don't you talk like that again!

MAID: Pardon, mistress, pardon! I was sent with such a message.

VASANTASENA: It is the message itself that I am angry with!

MAID: Then what shall I say to the mother?

VASANTASENA: Say this: 'If you wish that I should live, I should never again be asked to do such a thing.'

MAID: As you wish. (*exit*)

SARVILAKA (*entering*):—

Having cast on the night the slur of censure, having conquered sleep, and having got the better of the king's watchmen, here have I become devoid of lustre at sunrise, like the moon at the close of the night.

Moreover—

Whoever stares closely at me as I walk rapidly, or approaches me quickly as I stand in a state of distraction,—my suspicious mind internally treats all such persons equally; for it is one's own faults that fill a man with apprehension.

It was, indeed, for Madanika's sake that I undertook a risky adventure.—

In some place I had to let go a man who was busy talking to his servants; in another I passed by a house seeing that it was inhabited by women only; and sometimes I had to stand like a wooden pillar of a house when a posse of the king's watchmen came near me. By hundreds of acts, mostly like these, I turned the night into day.

(*Walks about*)

VASANTASENA: Girl, just take this painting-board and place it on my bed; then come back here quickly with a fan.

MADANIKA: As the mistress commands. (*exit, with the board*)

SARVILAKA: Here's Vasantasena's house. Well, then, let me go in. (*entering*) Where, possibly, can I find Madanika?

(*Enter* Madanika *with a fan in her hand*)

SARVILAKA (*seeing*): Ah! Here's Madanika!—

This is she, who, excelling even Madana[1] by the excellence of her charms, looks as if she were Rati[2] herself in bodily form. She renders my heart, heated by the fire of Love, exceedingly cooled as it were with sandal-paste.

O Madanika!

MADANIKA (*seeing*): Ah! What! Here's Sarvilaka! You are welcome, Sarvilaka! What brings you here?

SARVILAKA: I'll just tell you.

(*They lovingly cast glances at each other*)

VASANTASENA: Madanika has taken a long time to come; where, then, I wonder, could she have gone? (*looks through the window*) Ha! Here is she, engaged in talking with some man. Since she is gazing at him fixedly as if drinking him up with her steady and very loving glances, I guess that this must be that person who is desirous of freeing her from her bondage. Let them enjoy—let them! Let nobody suffer from interruption in love-making. I will not at all call her.

MADANIKA: Sarvilaka, tell it now.

(Sarvilaka *looks about him in all directions in fear*)

MADANIKA: Sarvilaka, what can this be? You seem to be afraid!

SARVILAKA: I am going to tell you some secret. So, is this spot quite secluded?

MADANIKA: Yes.

VASANTASENA: What, some deep secret! I shall not listen to it, then.

SARVILAKA: Madanika, will Vasantasena set you at liberty by accepting a ransom?

VASANTASENA: How! They are talking about me! Then I will listen to it through this window, keeping myself hidden from them.

MADANIKA: Sarvilaka, I did ask Her Ladyship, when she said: 'If I myself was free, I would set free all my servants from their bondage, even without any money, whatsoever.' So, Sarvilaka, how could you ever have so much money, so that you could liberate me from the service of Her Ladyship?

1. God of love in Hindu mythology.
2. Goddess of love in Hindu mythology.

SARVILAKA:—

Overpowered by poverty and led on by my love for you, O timid one, I committed a crime last night, all for your sake.

VASANTASENA: His appearance is pleasing to sight, but it inspires fear on account of his adventurous deed.

MADANIKA: Sarvilaka, for the sake of the trifling thing called woman, you have now placed both these in jeopardy—

SARVILAKA: What two things?

MADANIKA: Your life and your character.

SARVILAKA: You ignorant girl, fortune abides in adventure.

MADANIKA: Sarvilaka, you have hitherto been of blameless character. So may I hope that when for my sake you committed that crime, you did nothing which was very objectionable?

SARVILAKA:—

I do not rob a woman with ornaments on, looking like a creeper in blossom; I do not steal a Brahmana's wealth, nor gold collected for the purpose of a sacrifice; so, too, I never, during my search after money, rob a child lying on a nurse's lap. Thus my mind has always been given to the discrimination of what is right and wrong, even in the act of committing a theft.

So please request Vasantasena:—'Please wear in secret, out of love for me, this ornament which looks as if it had been specially manufactured to the exact measure of your body.'

MADANIKA: Sarvilaka, an ornament meant to adorn shall not be kept in secret. So bring it to me, please; I shall have a look at the ornament.[3]

SARVILAKA: Here's the ornament. (*gives it to her; hesitatingly*)

MADANIKA (*inspecting it*): This ornament looks like one seen before. Se tell me where you got it from?

SARVILAKA: Madanika, what have you to do with it? Take it.

MADANIKA (*angrily*): If you don't trust me, then why do you want to ransom me?

SARVILAKA: Well, I heard it in the morning in the merchants'

3. Courtesans are supposed to be decorated when they go out, not at home.

square that it belonged to the merchant Charudatta.

(*Both* Vasantasena *and* Madanika *act swooning*)

SARVILAKA: Madanika, compose yourself! Why do you now,

Quake and do not show compassion to me, when you are being freed from bondage, you, with all your limbs drooping in despair, and your eyes rolling in agitation!

MADANIKA (*recovering*): You rash fellow, I hope you didn't kill or wound anybody in that house, when you committed this foul deed for my sake?

SARVILAKA: O Madanika, Sarvilaka strikes not a person who is afraid, nor one who is asleep. So you may rest assured that I neither killed nor wounded anybody.

MADANIKA: Really?

SARVILAKA: Certainly.

VASANTASENA (*regaining consciousness*): Wonderful! I am restored to life again!

MADANIKA: It is pleasant news to me!

SARVILAKA (*jealously*): Madanika, what do you mean by saying it is pleasant news?—

Although born in a family with its ancestors all men of righteous conduct, still I committed a sinful deed, my heart being dominated by love for you. But, although my virtues have perished through my passion, still I do preserve my sense of honour. You call me a friend, and yet at the same time you run after another lover! (*woefully*)— In this world, the fine trees in the form of nobly-born youths having their whole wealth for their fruit, are completely divested of their fruit when devoured by the birds in the form of courtesans.—

And this fire of Love has amorous sports for its flames and affection for its fuel, wherein are offered as oblations men's youths and riches.

VASANTASENA (*smilingly*): Oh! His excitement is quite out of place.

SARVILAKA: In every way,—

Senseless are those men, I esteem, who trust in women or in riches; for riches as well as women move about as tortuously as female snakes.

Undue love should not be shown to women; women disrespect a man who is over-attached to them. Only that woman should be loved who is herself in love; but one whose heart is cold should be let alone.

It is indeed truly observed:—

'These harlots laugh, and weep too, for money; they make a man confide in them, but themselves never trust him. Therefore, a man of high birth and good character should avoid courtesans, like, jasmines in cremation-grounds.'—

'Of an unstable nature like the waves of the ocean, and exhibiting *raga* for a short time like the ridges of clouds in the evening, women, having taken a man's money, abandon him, when he is penniless, like a ball of *alaktaka*[4] with all its colour squeezed out.'

Indeed women, who are fickle,—

Having fixed upon one man in their heart, invite another by their glances; they pour an effusion of joy on a third one, while they carnally desire quite another still.

Indeed, it has been truly said by some one that—

'The lotus-plant grows not on mountain-top; asses bear not the yoke of horses; nor do grains of barley, when sown, yield a crop of rice; similarly, women born in the courtesans' quarter can never be pure.'

O you villain, accursed Charudatta! Here you cease to exist! (*walks a few steps*)

MADANIKA (*catching him by the hem of his cloak*): You utterer of nonsense, you have got angry at what is inconceivable.

SARVILAKA: How inconceivable, pray?

MADANIKA: This ornament, indeed, belongs to Her Ladyship.

SARVILAKA: What of that?

MADANIKA: And it was kept as a deposit with that gentleman.

SARVILAKA: For what purpose?

MADANIKA (*in his ear*): It was like this.

SARVILAKA (*with shame*): Ah! Alas!—

That branch of a tree, to which I had resorted for shade when

4. Red lac or a red paint prepared from the red resin-sap of certain trees, used by women for dyeing certain parts of their bodies, especially the feet and the lips.

suffering from summer's heat, that very branch has been despoiled of its foliage by me, in my ignorance.

VASANTASENA: How now! Even he is pained! So he must have done it in ignorance.

SARVILAKA: O Madanika, what is proper to be done now?

MADANIKA: As to that matter you yourself are clever enough to know.

SARVILAKA: Nay, say not so. See:—

These women, as is well-known, are forsooth clever by their very nature; while cleverness has to be taught to men by means of the *Sastras.*

MADANIKA: Sarvilaka, if you will listen to my advice, then restore this to that same high personage.

SARVILAKA: Madanika, what if he were to inform against me at the royal palace?

MADANIKA: Heat comes not from the Moon.

VASANTASENA: Well said, Madanika, well said!

SARVILAKA: O Madanika,—

For this my adventure I do not indeed feel sorrow nor fear: Why do you then narrate to me that good man's virtues? Does this ignoble deed produce any shame in me? And what can the king do to rogues like me?

Still, this is against policy. So, think of some other way.

MADANIKA: Then here's another way—

VASANTASENA: What may that other way possibly be?

MADANIKA: Be a person connected with that same gentleman, and take this ornament to Her Ladyship.

SARVILAKA: If I do so, what would come of it?

MADANIKA: In the first place, you will be no thief; while that gentleman, too, will be out of debt; and Her Ladyship would get back her own ornament.

SARVILAKA: Why, this is very risky!

MADANIKA: My dear fellow, take it to her; if you don't, *that* would be very risky!

VASANTASENA: Noble, Madanika, noble! You have spoken like a gentlewoman!

SARVILAKA:—

By following you, I have obtained excellent wisdom. When at night the moon is not visible, it is hard to find a guide who would point out the way.

MADANIKA: Then please wait for a short while in this Cupid's temple, while I shall announce your arrival to Her Ladyship.

SARVILAKA: So be it!

MADANIKA (*approaching*): Lady, here has come a Brahmana from Charudatta.

VASANTASENA: Girl, how do you know that he belongs to Charudatta?

MADANIKA: Lady, do I not know my own connections?

VASANTASENA (*to herself, shaking her head and smiling*): Properly said! (*aloud*) Let him come in.

MADANIKA: As Your Ladyship commands. (*approaching*) Sarvilaka, you may come in.

SARVILAKA (*approaches; with embarrassment*): Hail to Your Ladyship!

VASANTASENA: Sir, I salute you. Please be seated.

SARVILAKA: The merchant, Charudatta, sends this message: 'On account of the dilapidated state of my house, it is difficult to guard this casket. So please take it back.' (*he gives it to* Madanika, *and starts to go*)

VASANTASENA: Sir, you should carry back to him a counter-message from me, too.

SARVILAKA (*to himself*): Who is going there? (*aloud*) What is the counter-message?

VASANTASENA: Your Honour should take Madanika with you.

SARVILAKA: Lady, I do not understand this.

VASANTASENA: But I do.

SARVILAKA: What do you mean?

VASANTASENA: I was told by noble Charudatta that I was to give

Madanika to him who would hand over to me that ornament. So you may understand that he himself is bestowing her on you.

SARVILAKA (*to himself*): Ah! I have been found out by her! (*aloud*) Excellent, noble Charudatta, excellent!—

Men should always make an effort to acquire virtues. A virtuous person, poor though he may be, cannot be the equal of persons who are rich, but not virtuous.

Moreover—

An effort should be made by man to acquire virtues; for there is nothing that is exceedingly unattainable by means of them. Thus, the Moon, on account of the excellence of his merits, came to occupy Siva's head[5] which is otherwise inaccessible.

VASANTASENA: Is there any driver here?

(*Entering with a carriage*)

SERVANT: Your Ladyship, the carriage is ready!

VASANTASENA: Girl Madanika, look at me very well. You are given away. Mount the carriage; and remember me.

MADANIKA (*weeping*): Your Ladyship has abandoned me! (*she falls at her feet*)

VASANTASENA: Now you yourself have become worthy of salutation. So go and ascend the cart. And remember me.

SARVILAKA: Good bye, madam! O Madanika,—Take a good view of Her Ladyship and bow to her with bent head, from whom you have obtained the addition of the title of 'a bride' which is hard to obtain.

(*Mounts the carriage along with* Madanika, *and starts to go*)

5. According to Hindu mythology, Daksha cursed *Chandra* (Moon) that he would gradually fade away. Chandra didn't know what to do. He went and sought advice from Brahma who told him that the only recourse was to pray to Shiva. Chandra made a linga on the banks of the river Sarasvati and meditated. At the end of the *tapasya* Shiva appeared before Chandra and offered to grant him a boon. He prayed to Shiva for six months. Lord Shiva wore the moon crescent on his head, making him grow for 15 days and decay for 15 days periodically. Charudatta's virtuosity is compared with the moon, here.

(*Offstage*)

Who is there? Listen, The chief of the police orders: 'King Palaka, alarmed on account of his belief in the prophecy of Siddha that Aryaka, a cowherd's son, would be the king, has now brought him from his hamlet and confined him in a dreadful dungeon. So you should all watch carefully at your respective stations.'

SARVILAKA (*listening*): What! King Palaka has arrested my dear friend Aryaka; and here I have just got a wife! Oh! Bad luck! Or rather—

In this world, these two things are exceedingly dear to men; but now the circumstances are such that the friend has become more important than even a hundred fair wives.

All right. I'll get down. (*does so*)

MADANIKA (*tearfully; folding her hands*): It is just so. Well, then let my lord send me now to my elders.

SARVILAKA: Good, my dear, well said! You have spoken just what is in accord with my mind! (*to the servant*) My good fellow, do you know the house of Rebhila, the merchant?

SERVANT: Yes.

SARVILAKA: Take my wife there.

SERVANT: As Your Honour orders.

MADANIKA: As my lord says. But my lord should be careful. (*exit*)

SARVILAKA: Now I will—

Incite to rebellion, mine and Aryaka's kinsmen, the city rakes, men that have won fame through the valour of their own arms, and such servants of the king as have been alienated through being insulted by him,—all for the release of my friend, just as Yaugandharayana did incite people for the release of King Udayana.[6]

6. Sarvilaka is comparing himself with Yaugandharayana, the faithful minister of King Udayana (as Sarvilaka to Aryaka here), instrumental in setting him free using a clever stratagem. Udayana was a king of the lunar race, the son of Sahasranika and the grandson of Satanika (son of Janamejaya) who transferred the capital of upper India from Hastinapura to Kausambi. He was so named because he was educated on the Eastern (Udaya) Mountain by the

Moreover—

My dear friend, who has been arrested causelessly by the wicked enemies, feeling alarmed in their minds, I will release by a sudden onset, as if he were the lunar orb lying in the jaws of Rahu[7]. (*exit*)

SERVANT (*entering*): Lady, fortune smiles on you! A Brahmana has come from the noble Charudatta.

VASANTAṢENA: Oh! How charming is this day! So, girl, bring him in with due respect, attended by one of the bastard-pages.

MAID: As Your Ladyship commands. (*exit*)

(*Enter* Vidushaka, *attended by* a *bastard-page*)

VIDUSHAKA: Oh! Wonder! Ravana, the king of the Rakshasas, used to travel by the aerial car, Pushpaka, which he had conquered by means of the pain he suffered in practising penance. But here I, a Brahmana, am moving about, attended by a number of men and women, although I have suffered no pain.[8]

---

sage Jamadagni. Mahasena, also called Chanda-Mahasena, had a daughter, named Vasavadatta. She once saw Udayana in a dream, and being enamoured by him contrived to inform him of her love (subject matter of Bhasa's *Svapnavasadatta*). Udayana was decoyed to Ujjayini by Mahasena, and was engaged to teach music to Vasavadatta, managed to elope with her to his capital, following the advice of his minister Yaugandharayana. They were afterwards married with great pomp.

7. When *Amrita* (ambrosia) was being served to the gods after it was churned from the ocean, Rahu, disguised as a god, joined the rank of the gods. Nectar was poured out to him; but the fraud was detected by the Sun and the Moon who informed Vishnu. The latter severed the head of the demon just as he was swallowing the nectar. As he had tasted a little of the quantity of nectar, the head became immortal, and it is supposed to wreak its vengeance on the Sun and the Moon. Astronomically, Rahu is the ascending node personified as a demon. He is also considered as one of the planets, as King of Meteors and guardian of the Southwest quarter. The significance here is that Sarvilaka's friend, Aryaka was under a cloud, like the moon in the grasp of Rahu.

8. Reference from the epic *Ramayana*. Ravana became invincible to the gods after he had practised severe austerities and proficiated Brahma. Vidushaka thinks that he is happier as he is being attended upon by people without having had to practise rigid penance for it. *Pushpaka* is the aerial cart of Ravana which he acquired from Kubera.

MAID: Sir, please look at this entrance to our abode.

VIDUSHAKA (*looks, admiringly*): Oh! How beautiful is this gate of Vasantasena's mansion! It is besprinkled with waters swept, and then besmeared with greenish cowdung. Its floor space is attractively decorated with the offerings of various kinds of sweet-smelling flowers. It has raised its top to a great height, as if out of curiosity to have a sight of the celestial regions. It is adorned with a string of a huge garland of jasmine flowers which is tossing about and hanging down from it, making one mistake it for the trunk of Airavata[9]. It is splendidly decked with a lofty arch of ivory. It is decorated with a number of auspicious banners, which look lovely with the lustre of costly jewels, whose finger-like tips are tremulously shaking as they are tossed about by the force of the wind, and which thus seem to be as it were beckoning me to get in. Both its sides are rendered attractive by auspicious crystal-jars which are kept on the pedestals of the pillars supporting the archway, and which are charming on account of the green mango-leaves placed gracefully therein. Its golden panels are thickly set with diamonds, and are as hard to break open as the chest of a big demon. It tortures the desires of indigent persons. To tell the truth, it forcibly arrests the eye even of an indifferent looker-on like me.

MAID: Come, come this way; please enter this first quadrangle.

VIDUSHAKA (*entering and looking about*): Oh! Wonderful! Here, too, in this first quadrangle, are rows of balconies resembling in colour the moon, or conches, or lotus-stalks, which are whitened with handfuls of chunam-powder laid over them as paint, which are adorned with golden stairs inlaid with various kinds of gems, and which seem to be, as it were, gazing down fixedly on Ujjayini with their moon-like faces in the form of the crystal-windows from which pearlstrings are dangling. Here the door-keeper is comfortably sleeping, lying at ease like a learned Brahmana! The crows, though enticed by means of *Kalama*[10] rice mixed with curds, do not eat the oblations, they being of the same colour as chunam-paint[11] and therefore indistinguishable from it— Please direct me further, madam.

9. Airavata is Indra's elephant. The festoon is very thick and white and waving. It looks like the trunk of Airavata, which is also white.
10. Rice which is sown in May-June and ripens in December-January.
11. White paint.

MAID: Come, sir, come this way; please enter this second quadrangle.

VIDUSHAKA (*entering and looking about*): Oh! Wonderful! Here, too, in this second quadrangle, there are tied the carriage-oxen, fattened on the mouthfuls of grass and corn-husks that are placed before them, and having their horns anointed with oil. Here's one buffalo, snorting like a high-born man when slighted. Here they are rubbing the neck of a ram as if he were a wrestler, after the fight is over. And here in this direction they are dressing the hair of some horses. Here's a monkey, fastened in its stable securely like a thief. (*looking in another direction*) And here an elephant is being fed by the keepers with balls of boiled rice mixed with ghee.—Please direct me further, madam.

MAID: Come, sir, come this way; please enter this third quadrangle.

VIDUSHAKA (*entering and looking about*): Oh! Wonderful! Here, too, in this third quadrangle, are these seats, duly arranged, for the use of well-born youths. There is a half-read volume lying on a gaming-table. Here's this gaming-table with chessmen made of precious stones. And here are these courtesans and aged gallants, wandering in all directions, who are skilled in managing love-unions and love-quarrels, carrying in their hands picture-boards covered with paintings in various colours.—Please direct me further, madam.

MAID: Come, sir, come this way; please enter this fourth quadrangle.

VIDUSHAKA (*entering and looking about*): Oh! Wonderful! Here, too, in this fourth quadrangle, the tabors struck by the hands of ladies are sounding deeply like clouds; cymbals[12] of bell-metal beating time are falling with a flash, like the stars from heaven when their merit is exhausted; and bamboo-pipes are being sweetly played upon, sounding like the hum of bees. Here's a lute which, after being placed on the lap like a maiden angered in jealous love, is being played upon with the touch of the nails. Here are some courtesans singing very sweetly, like female bees intoxicated with the honey of flowers, who are being taught to dance and made to learn a drama full of erotic sentiment. And water-jars, turned upside down, are kept suspended in the windows to

12. A common percussion instrument.

receive the cool breeze.—Please direct me further, madam.

MAID: Come, sir, come this way; please enter this fifth quadrangle.

VIDUSHAKA (*entering and looking about*): Oh! Wonderful! Here, too, in this fifth quadrangle, there pervades an overpowering odour of asafetida[13] and oil, exciting the appetite of a poor person. The kitchen, being always heated, is sighing, as it were, through its mouths in the shape of its doors, sending forth effusions of fragrant smoke of different kinds. The smell of the many varieties of edibles and condiments that are being prepared is exciting my appetite very much. Here is this butcher's boy washing the entrails of a killed animal, looking like a piece of an old garment. The cook is preparing various kinds of different eatables: *modakas*[14] are being made and *apoopas*[15] fried. (*to himself*) Oh, how I long I should be offered here water for my feet, with a request to eat to my heart's content! (*looking in another direction*) To tell the truth, this mansion in this spot, with its bastards and courtesan-folk adorned with different kinds of ornaments, looks as if it were heaven itself, with its crowds of Gandharvas and Apsarasas—Sirs, who are you that go by the name of '*Bandhulas*'?

BASTARDS: We are, indeed,—*Bandhulas*, that sport about like the cubs of elephants, being reared in other people's houses, fed on others' food, begotten by other men upon stranger-women, enjoying others' riches, and possessing no particular merits to speak of.[16]

VIDUSHAKA: Please direct me further, madam.

MAID: Come, sir, come this way. Please enter this sixth quadrangle.

VIDUSHAKA (*entering and looking about*): Oh! Wonderful! Here, too, in this sixth quadrangle, these architectural arches, made of jewels set in gold, and embossed on a background of sapphires, exhibit as it were a spot full of rainbows. Goldsmiths are consulting one another as

13. A fetid resinous gum obtained from the roots of a herbaceous plant, used in herbal medicine and Indian cooking.
14. A sweet dumpling popular in Western, Eastern and Southern India.
15. A sweet dish made with rice flour, jaggery and ghee.
16. It is considered to be an interpolation on the ground that the bastard pages were not expected to speak of themselves so disparagingly. But looking at the state of the society then and the position of harlots in it, it was no disparagement for bastards to speak of themselves in this fashion.

regards various kinds of jewels, such as lapis lazuli, pearls, corals, topazes, sapphires, cat's-eyes, rubies, emeralds, etc. Rubies are being set in gold; golden ornaments are being manufactured; pearl-ornaments are being strung on threads of red silk; lapis lazuli are being gently polished; conches are being cut; and corals are being ground on whetting stones. Layers of wet saffron-paint are exposed for drying; musk is being moistened; sandal-juice is being made up. Betel-rolls with camphor are being offered to the lovers of the courtesans; side-glances are being cast; laughter is going on; and wine is being incessantly drunk with sounds of merriment. Here are drinking these men-servants, these maid-servants, and also these men who have forsaken their children, wives and property, and are left to themselves by the courtesans who have gone away, having drunk wine from liquor-cups.—Please direct me further, madam.

MAID: Come, sir, come this way. Please enter this seventh quadrangle.

VIDUSHAKA (*entering and looking about*): Oh! Wonderful! Here, too, in this seventh quadrangle, pairs of pigeons are enjoying happiness, perched comfortably in snug dove-cots and busy in kissing each other. The parrot in its cage, fed on curds and rice, is reciting a Vedic hymn, like a Brahmana whose belly is filled with rice and curds.

And this talking thrush is chattering lustily like a house-maid that has gained ascendancy on account of the consideration shown to her. The cuckoo is cooing like a procuress, with her throat delighted by having tasted the juice of numerous fruits. Rows of cages are suspended from pegs. Quails are incited to fight; partridges are being made to talk; caged pigeons are being set upon each other while this domestic peacock, looking as if decked with various kinds of gems, is, as it were, fanning the whole mansion heated with the sun's rays, with the flutter of his tail, as he dances about joyfully hither and thither. (*looking in another direction*) Here these pairs of royal swans, looking as if they were the moon's rays rolled up, are walking behind lovely damsels, as though learning their gait. And here these tame cranes are moving here and there, like aged eunuchs. Oh! Wonderful! An extensive scene is presented by the courtesan's collection of various kinds of birds! To tell the truth, this mansion of the courtesan appears to me like the

Nandana garden of Indra![17]—Please direct me further, madam.

MAID: Come, sir. Please enter this eighth quadrangle.

VIDUSHAKA (*entering and looking about*): Girl, who is this person clad in a silken cloak and gorgeously decked with a superfluity of very wonderful ornaments, moving about haltingly and with distortions of the body?

MAID: Sir, this is Her Ladyship's brother.

VIDUSHAKA: How much penance has a person to practise to become Vasantasena's brother? Or rather,—

I must not think highly of him. For, although he is gaily dressed, gentle, and well-perfumed, still he is to be shunned by the people, just like a *Champaka*[18] tree growing in the enclosure of a cremation-ground, although it may be bright with blossoms, attractive, and sweet-smelling.

(*looking in another direction*) Girl, and who is this woman occupying a lofty seat, clad in a garment embroidered with flowers with her oil-greasy feet thrust into a pair of shoes?

MAID: Sir, she is the mother of our mistress.

VIDUSHAKA: Oh! How great is the expanse of the belly of this ugly she-ghoul![19] What, then, was the beautiful entrance of this house constructed after she was set up here, as is done with a huge idol of Siva?

MAID: You knave, you mustn't jeer at our mother in this fashion. She is suffering from a quartan fever.

VIDUSHAKA (*with a burst of laughter*): Holy quartan fever, look you also upon me, a Brahmana, with such a favour!

MAID: You wretch, you will die then!

VIDUSHAKA (*jestingly*): You slave-born wench, better is a fat and bloated belly like this dead!—

17. According to Hindu mythology, Nandana is a garden in devalok (celestial world) owned by Indra.
18. *Magnolia champaca*, Champaka in English, is a large evergreen tree, known for its fragrant flowers and its timber used in woodwork.
19. Ghoul is an evil spirit or demon. She-ghoul, here, is a reference by Vidushaka to Vasantasena's mother.

The mother has been reduced to such a condition, having intoxicated herself with spirits, liquors and wines. If the mother were now to die, she would provide sufficient food for a thousand jackals! Girl, have you any sailing vessels going abroad for trade?

MAID: Sir, no, not at all.

VIDUSHAKA: Or rather, why ask a question in this matter? For surely, in this ocean of your passion, having affection for its pure waters, your breasts, hips and buttocks are themselves the charming sailing vessels! To tell the truth, having thus seen this mansion of Vasantasena with its eight quadrangles and its numerous appurtenances, I feel I have, as it were, seen the three worlds brought together in one spot. I have not the power of speech to praise it, as it deserves, whether it should be called a courtesan's house, or a portion of Kubera's[20] palace. Where is your mistress?

MAID: Sir, here she is, in this garden. May Your Honour enter.

VIDUSHAKA (*entering and looking about*): Oh! Wonderful! How beautiful is this garden! Numerous trees have been planted here, that have put forth ample blossoms in a very charming manner. There is a silken swing, hung, beneath thickly-growing trees, and big enough to accommodate ladies' hips. To tell the truth, the garden throws into the background as it were the beauty of Indra's Nandana garden, by means of the flowers that have spontaneously dropped down, of the *Champaka*, the *Yuthika*, the *Sephalika*, the *Malati*, the *Mallika*, the *Navamalika*, the *Kurabaka*, the *Atimukta*[21] and other plants. (*looking in another direction)* And here this oblong well looks like red twilight as it were, owing to the lilies and the red lotuses blooming therein, whose lustre resembles that of the rising sun. Moreover—

Here this Asoka tree, having recently put forth blossoms and foliage, looks like a brave warrior in the thick of the battle, amply splashed all over with clotted blood.

Well, where is your mistress?

MAID: Sir, bring down your eyes, and you'll see Her Ladyship.

20. Lord of riches in Hindu mythology.

21. *Yuthika* is a kind of jasmine; *Sephalika* is a fragrant white flower; *Malati* and *Mallika* are kinds of jasmine; *Navmalika* is a double jasmine; *Kurabaka* is red amaranth; and *Atimukta* is the spring creeper.

VIDUSHAKA (*seeing her and approaching*): My greetings, madam!

VASANTASENA (*speaking in Sanskrit*)[22]: Oh! Here's Maitreya! (*getting up*) Welcome sir! Here's a seat; pray be seated here.

VIDUSHAKA: You too may sit down.

(*Both sit down*)

VASANTASENA: Is the merchant's son well?

VIDUSHAKA: Madam, he is well.

VASANTASENA: Sir, Maitreya, do now—

The birds in the form of friends happily resort to tha tree in the form of a good man? The tree which has accomplishments for its foliage, modesty for its principal branches, popular trust for its root, and fame for its flowers, and which is richly endowed with fruits in the form of merits?

VIDUSHAKA (*to himself*): The roguish blonde has observed acutely! (*aloud*) Yes, so they do.

VASANTASENA: Now, what may be the object of your visit?

VIDUSHAKA: May madam listen! His Honour Charudatta, placing his folded hands on his head, begs to say—

VASANTASENA (*folding her hands*): What does he command?

VIDUSHAKA: 'I lost that golden casket at gambling, in the confident belief that it belonged to me. And I do not know where that gambling-master, the king's emissary, is gone.'

MAID: I congratulate you, madam! His Honour has turned a gambler!

VASANTASENA (*to herself*): How! Although it was stolen by a thief, still he says, out of his magnanimity, that it was lost at play! Hence it is that I love him!

VIDUSHAKA: 'So in exchange for that, Your Ladyship may kindly accept this jewel-necklace.'

VASANTASENA (*to herself*): Shall I show him that ornament? (*considering*) Or rather, not so soon.

VIDUSHAKA: What then? Is not Your Ladyship disposed to accept this jewel-necklace?

22. For details see D. R. Ankur's essay.

VASANTASENA (*laughs, looking at the face of her* Maid): O Maitreya, how shall I not accept the jewel-necklace (*taking it, keeps it by her side; to herself*) How! Even from the mango-tree denuded of all blossoms there still do fall drops of honey! (*aloud*) Sir, please tell the worthy Charudatta, the gambler, in my name, that I too shall be visiting him this evening.

VIDUSHAKA (*to himself*): What other thing will she get by going there? (*aloud*) Lady, I will tell him. (*to himself*)—That he should refrain from having anything to do with this courtesan. (*exit*)

VASANTASENA: Girl, keep this ornament. We shall now go to sport in the company of Charudatta.

MAID: Lady, look, look here; an untimely storm is gathering in the sky.

VASANTASENA:—

Let clouds gather as they will; let it turn night; or let rain fall incessantly. With my heart yearning for my beloved, I care not for all of them.

Girl, take a garland and come quickly.

(*Exit all*)

End of Act IV, entitled

'Madanika and Sarvilaka'

Act V

(*Enter seated love-sick* Charudatta).

CHARUDATTA (*looking upward*): An untimely storm is gathering. For here this,—

Untimely storm, which is joyfully gazed upon by the tame peacocks with their uplifted tails, and is disliked by the distressed swans who are about to depart to the Manasa lake,[1] quickly pervades the sky, and also afflicts at the same time the heart of a lover.

Moreover—

The cloud, looking dark like a bee or like the belly of a buffalo wet with water, wearing an upper garment of yellow silk constituted by the flash of lightning, and bearing conches in the form of serried flocks of cranes, is about to traverse the sky, as though it were a second Vishnu[2].

Moreover—

There rises forth a cloud, dark like the body of Vishnu, forming an appearance of conches by means of the crooked lines of cranes, and having the yellow silken mantle in the form of the streaks of lightning,— as though it were Vishnu[3] himself.—

Here these streams of water, resembling downpours of molten silver, dropped rapidly from the interiors of clouds, fall, like the rent hems[4] of the garment of the sky, seen for a moment in the light of the

1. It is believed that the lake was first created in the mind of Brahma before he manifested it on Earth. Hence it is called 'Manasa sarovaram', which is a combination of the Sanskrit words for "mind" and "lake". Manasarovar or Manasa Lake is a poetic convention, located in Himalaya mountains, where the *hansa* (wild grey goose or swans) migrate annually.

   The *hansa* is a mythical or poetical bird with knowledge. Considered to be sacred, it is an important element in the symbology of the subcontinent, representing wisdom and beauty. In the *Rig Veda*, this bird is able to separate Soma (ambrosia) from water, when mixed. In the post-Vedic literature, it is this same bird which separates milk from water.
2. Vishnu as dark, who wears an upper garment of yellow silk, and has in his hand a white conch.
3. Here, the discus-bearer.
4. Fringes of the vast cloth of the sky, lit up by the flash of lightning.

flame of the lamp of lightning, and then lost to view.—

This sky looks as if it had been adorned with decorative paintings of leafy lines, owing to the high-roaring clouds, scattered by the wind, and assuming various expansive shapes, some resembling thronging couples of *Chakravaka*[5] birds, some flying swans, some alligators and fish-swarms thrown about pell-mell, and some lofty mansions.—

Here is this sky, darkened with clouds, that resembles the army of the Kauravas, producing darkness. This peaçock, greatly flushed with the pride of his vigour, is shouting in glee, like Duryodhana, whose army was mightily proud. The cuckoo has ceased cooing, like Yudhishthira taking to the path of the forest, when conquered at dice-play. And, on account of the muddy water, the swans have now gone off to dwell in an unknown place, like the Pandavas living incognito after leaving the forest.[6]—

(*reflecting*) It is a long time since Maitreya went to Vasantasena. How is it that he does not come even now?

(*Entering*)

VIDUSHAKA: Oh, the greed and the want of courtesy of that courtesan, that she did not even talk about anything else! She spoke variously about something in a manner befitting her affection and thus accepted the jewel-necklace. With all that wealth she did not say to me: 'Sir Maitreya, rest yourself here; you may depart after you have drunk water at least, in a cup.' Never then shall I so much as even look at the face of that slave-born harlot! (*sorrowfully*) Truly do they say: 'A lotus-plant that grows without a root; a tradesman that does not cheat; a goldsmith that doesn't steal; a village-meeting that is without a quarrel; and a courtesan that is not avaricious;—these are hard to be found.' So I'll go to my friend and dissuade him from this his

5. Legend has it that lovers who had committed the 'sin' of disturbing sages at their meditation were cursed to turn into Chakravaka birds (ducks) in their next birth: whence, the pair will have to separate after dusk and would, therefore, be reduced to calling one another piteously through the night.

6. The *Mahabharata's* account of Yudhishthira's gambling away his kingdom is well known. According to conditions imposed upon the Pandavas by Duryodhana (Kaurava prince), they had to lead a solitary life for twelve years in a forest and remain incognito for one year. Vis-ā-vis the preceding reference, the swans, too, pass secretly to unknown regions (Manasa Lake).

attachment to that courtesan. (*walking about and observing*) How! Here's my dear friend sitting in the garden. Well, I'll approach him. (*approaching*) My greetings to you! May you prosper!

CHARUDATTA (*seeing*): Ah! My friend Maitreya has come! Welcome, friend! Pray be seated.

VIDUSHAKA: I have taken my seat.

CHARUDATTA: Friend, tell me about that business.

VIDUSHAKA: Well, that business has failed!

CHARUDATTA: What, did she not accept that jewel-necklace?

VIDUSHAKA: Whence could we have so much luck! She took it, placing on her head her folded hands, delicate like fresh water-lilies.

CHARUDATTA: Then how do you say that the business failed?

VIDUSHAKA: Friend, how has it not failed, since you have given away the jewel-necklace, the very essence as it were of the four oceans, in exchange for a golden casket of little value, that we neither ate nor drank, and that was stolen by thieves?

CHARUDATTA: Nay, friend, say not so!—

This price is being paid, not for that ornament, but for that high trust, by relying on which she kept that deposit with us.

VIDUSHAKA: O friend, I have this another reason, too, for being angry, inasmuch as she made a sign to her maid, and, veiling her face with the end of her garment, she laughed derisively at me. So I, being a Brahmana, now bow to you with my bent head and request you that you should turn yourself away from this affair with that courtesan, which is full of many pitfalls. A courtesan is just like a pebble got inside a shoe, that is got rid of afterwards with great difficulty. Moreover, O friend, a courtesan, an elephant, a Kayastha[7], a mendicant, a spy and a donkey,—where these dwell, there not even villains can flourish.

CHARUDATTA: Friend, now you need no longer give vent to all these strictures; I am warded off from that pursuit by my condition itself. See—

A wearied horse puts on speed to go quickly, but his legs do not move to that extent for want of vigour. Similarly, the volatile desires

7. A scribe.

of a man wander everywhere, but, weary at failure, they afterwards enter once more in his bosom.

Moreover, friend,—

Whoever has riches, will have that lady; for she is to be bought with money.

(*to himself*) No, she is surely to be bought with merits!

(*aloud*)—But riches have left me; she therefore is, of course, left by me.

VIDUSHAKA (*looking down; to himself*): Since he is looking upward and sighing deeply, I conclude that his passion must have become still more inflamed as I tried to dissuade him. So it has been truly said that 'Love is perverse.' (*aloud*) O friend, she also said: 'Tell Charudatta that I shall visit him there this evening.' So I guess that not being satisfied with that jewel-necklace, she is coming to demand something in addition.

CHARUDATTA: Friend, let her come. She will go satisfied.

CHETA (*entering*): Listen, all people!—

The more the bank of clouds rains, the more wet does the skin of my back become; and the more the cold wind beats, the more does my heart quake. (*laughing loudly*)—

I can play upon the loud-sounding flute with seven holes and upon the resounding lute with seven strings. I can sing songs just like an ass. What is Tumburu or Narada[8] as compared to me, in singing?

I am ordered by Her Ladyship Vasantasena: 'Kumbhilaka, go and inform the worthy Charudatta of my arrival'. So, I'll go to the house of the noble Charudatta. (*walks about; entering and seeing*) Here is Charudatta, seated in his garden. Here, too, is that rascal. Well, then, I'll approach them. How! The gate of the garden is closed! Never mind, I'll make a sign to this rascal. (*throws lumps of mud*)

VIDUSHAKA: Ah! Who is this here that is now pelting me with clods, as if I were a *Kapittha*[9] tree enclosed within a fence?

CHARUDATTA: Perhaps a clod might have been let fall by the

8. Narada, son of Brahma, and Tumburu, the Gandharva with a horse's head, are described as the best of the celestial musicians.
9. Wood apple or the Elephant apple.

pigeons playing on the terrace of the garden-house.

VIDUSHAKA: You slave-born wretch of a pigeon, you just wait till with this wooden staff I make you fall off on the ground from this house, as if you were a ripe mango-fruit ! (*lifts his wooden staff and runs*)

CHARUDATTA (*holding him by his sacred thread*): Sit down, friend; why do you bother yourself with him? Leave that poor pigeon alone, who is in the company of his mate.

CHETA: How! He is looking at the pigeon, and not at me! Well, I'll again strike him with another lump of mud. (*does so*)

VIDUSHAKA (*looking in all directions*): Ah! It is Kumbhilaka! Well then, I'll go to him. (*approaching and opening the door*) Well, Kumbhilaka, come in. You are welcome.

CHETA (*entering*): Sir, I salute you!

VIDUSHAKA: Where have you come, man, on such a rainy day, and in darkness?

CHETA: You see, she is here.

VIDUSHAKA: What 'she'? What 'she'?

CHETA: *She*. See? *She*!

VIDUSHAKA: You son-of-a-slave, what makes you now hiss forth 'she-see-she,' like an old beggar in famine-time?

CHETA: Eh, why do you too now shout out: 'what, what', 'ka, ka', regularly like a greedy crow crowing during Indra's festival?

VIDUSHAKA: All right. Tell me.

CHETA (*to himself*): Well, I'll speak thus. (*aloud*) Eh, I'll set a puzzle to you!

VIDUSHAKA: And I'll set my foot on your head!

CHETA: Oh, all right! Then just find out in what season the mango-trees blossom.

VIDUSHAKA: You son-of-a-slave, in the *Grishma*[10]!

CHETA (*laughing*): Ah, no! no!

VIDUSHAKA (*to himself*): What answer then should I make here? (*pondering*) Well, I'll go to Charudatta and ask him. (*aloud*) Eh, you

10. Summer season

wait awhile. (*approaching* Charudatta) O friend, let me just ask you in what season the mango trees blossom.

CHARUDATTA: In *Vasanta,* you fool!

VIDUSHAKA (*going to* Cheta): In *Vasanta,* you fool!

CHETA: I'll set you another puzzle: Who affords protection to prosperous villages?

VIDUSHAKA: Eh, *Rathya*[11].

CHETA (*laughs*): Ah, no! no!

VIDUSHAKA: Oh! well, I have fallen into doubt. (*pondering*). Never mind; I'll again ask Charudatta. (*again goes back and asks* Charudatta *as before*).

CHARUDATTA: Friend, *Sena*[12].

VIDUSHAKA (*approaching* Cheta): You son-of-a-slave, *Ṣena.*

CHETA: Now put the two together and repeat it quickly.

VIDUSHAKA: Sena-Vasante![13]

CHETA: No, say it turned around.

VIDUSHAKA (*bodily turning around*): Sena-Vasante!

CHETA: You foolish fellow, turn the *padas*[14]!

VIDUSHAKA (*turning his padas*): Sena-Vasante!

CHETA: You idiot, turn the *padas,* formed by those letters!

VIDUSHAKA (*pondering*): Vasanta-Sena!

CHETA: Here she has arrived!

VIDUSHAKA: Well, I'll just go and tell Charudatta. (*approaching*) Sir Charudatta, a *Dhanika* has come to see you!

CHARUDATTA: Why should a *Dhanika*[15] visit my house?

11. A collections of *rathas* or carts bringing corn to the city.
12. The army.
13. Vidushaka purposely puts the words in wrong order, so that there should be room for some quibbling. Maitreya, quibbling on the word, repeats the same thing again.
14. Words; also meaning feet.
15. Literally, a creditor or a rich person. Vidushaka humoursly makes a pun by calling Vasantasena thus. It would mean she has come to claim money or to give.

VIDUSHAKA: If not in your house, the *Dhanika* is at your door. Here's Vasantasena come to see you!

CHARUDATTA: Why do you deceive me friend?

VIDUSHAKA: If you don't believe what I say, then ask this Kumbhilaka: You son-of-a-slave, Kumbhilaka, come here!

CHETA (*approaching*): Sir, I salute you!

CHARUDATTA: Welcome, my good man, tell me, has Vasantasena *really* come?

CHETA: Here she has come, our Vasantasena.

CHARUDATTA (*joyfully*): My good fellow, I have never let a welcome announcement go by without bestowing a reward. So take this reward. (*gives his upper garment*)

CHETA (*takes it and bows, with delight*) Well, I'll tell Her Ladyship. (*exit*)

VIDUSHAKA: Sir, do you know for what purpose she has come on such a rainy day?

CHARUDATTA: Friend, I can't quite understand it.

VIDUSHAKA: But I do. Not being satisfied, as the necklace is of small value and her golden casket more valuable she has come to demand something in addition!

CHARUDATTA (*to himself*): She shall go away fully satisfied.

(*Enter love-sick* Vasantasena *brilliantly attired as an Abhisarika*[16], *accompanied by her* Umbrella-bearer *and by* Vita)

VITA (*with reference to* Vasantasena):—

She is the goddess Lakshmi, without the lotus in her hand; she is the lovely weapon of Cupid; she is the cause of grief to high-born chaste ladies; and she is the flower of the excellent tree in the form of Love. Exhibiting bashfulness at the time of amorous sports, she is followed by hosts of gallants wandering after her, as she walks gracefully over the stage in the form of Love's field.

See, Vasantasena, see!—

With their forms hanging low on the peaks of mountains, and resembling in colour the darkened hearts of ladies in separation, the

16. See Glossary.

clouds are thundering, whose roar makes the peacocks start up suddenly, so that the sky is fanned by their opened tails as though by jewelled fans.

And also—

The frogs, pelted by rain-showers, and with their mouths besmirched with mud, are drinking water; the impassioned peacock lets loose his voice; the *nipa*[17] tree, with its red blossoms, glows like a bright lamp; the clouds have obscured the moon, like rakes taking to the life of a *Samnyasin*[18]; while the lightning stays not in one spot, like a woman born of a low family.

VASANTASENA: Sir, you have described correctly. See, here this—

Co-wife in the form of the night is blocking my path in anger, forbidding me now and again with its thundering, as though saying— You foolish woman, what business have you here when your lover sports with me alone, possessing plump *payodharas*[19]?

VITA: Yes, it is as you say. You may therefore scold her.

VASANTASENA: Sir, what is the use of scolding her who is obstinate, as is women's nature? And sir, observe—

Let the clouds pour in torrents, thunder or shoot forth lightning; women who go to meet their lovers do not mind cold or heat.

VITA: See, Vasantasena, see! This other—

Cloud, having its speed augmented by the breeze, discharging a stream of arrows in the form of big torrents, with its rumbling serving as the beat of drums, and displaying banners in the form of lightning, seizes the cluster of the moon's rays, in the sky, like a king receiving from an enemy, weak in valour, large tributes in the latter's capital; a king having movements rapid like those of the wind, discharging

17. The Nipa, also known as Kadamba, has bright buds and blossoms at the roaring thunder, when sprinkled with fresh rainwater.
18. People who lead a dissolute life readily become samnyasins to cover their vile deeds.
19. The commentary on the play suggests that Vasantasena calls the 'night' her 'rival-wife', having clouds (breasts of the night) so close that they touch each other, leaving no space between. There is nothing to support such a supposition except the quibbling on *payodhara*.

streams of arrows like big torrents, with sounds of drums like the rumbling of clouds, and displaying banners like lightning.

VASANTASENA: So it is. But then how is it that here this another—

Wicked crane, with villainous intentions, and serving as the drum that sounds the doom of ladies whose husbands have gone on a journey, is also crying '*Pravrit, Pravrit*'[20], sprinkling salt on their wounds as it were, when their minds are already tortured by these thundering clouds dark like big elephants, with their interiors swollen and hanging low, and variegated with cranes and with lightning-flashes?

VITA: So it is, O Vasantasena! But see this another thing:—

The sky, having a white headgear in the form of the cranes, and waving *chowries* in the form of lightning-flashes, is, as it were, desirous of imitating the appearance of an infuriated elephant.

VASANTASENA: See, sir, see!—

By these clouds, dark like the wet leaves of the *tamala*[21] tree, the sun has been completely hidden in the sky; the anthills, beaten with the torrential rains, crumble down, like elephants cowering down when struck by arrows; lightning is made to move through the mansions like a golden lamp; and the moon-light, like the wife of an imbecile husband, has been taken away after being forcibly seized.

VITA: See, Vasantasena, see!—

These clouds pouring torrents of rain with their sides girt by streaks of lightning, and rushing against each other, like elephants round whose armpits are fastened lightning—like chains made of gold, are, as it were, lifting up the earth with silver chains, at the command of Indra, their lord.

And see also—

By the clouds, thundering in the strong gale, dark like a herd of buffaloes, and moving about like oceans in the sky on their wings in

20. Literally, 'Rain, Rain.' The hearts of the women were already wounded; the cry of the crane, which constantly reminded them of the rainy reason, made their case worse and thus acted like salt cast on a wound.

21. *Cinnamomum tamala* or *Tejpat*, Malabar leaf, Indian bark known for its medicinal value; the bark of a tamala tree is also used for cooking.

the form of lightning, this earth, emitting a strong odour, and covered with the blades of new green grass, is pierced with the sharp falling showers, as if with arrows made of jewels.

VASANTASENA: Sir, here this another—

Cloud is rising forth, darkening the quarters as if with black collyrium[22], very loudly welcomed by the cries of peacocks shouting 'come, come'; ardently and lovingly embraced, as it were, by the cranes flying up; and looked at with very great distress by the swans that have left the lotus-beds.

VITA: So it is. And also mark:—

The world is now sleeping motionless, as it were, lying in the inside of the mansion formed by the torrents of water; its eyes in the form of lotus-clusters are rendered motionless; it has lost all consciousness of night and day; owing to the lightning-flashes, the darkness therein momentarily disappears and becomes visible again; its face in the form of the quarters is veiled; and it has a covering canopy of numerous clouds in the spacious sky, the abode of clouds.

VASANTASENA: Sir, it is as you say; and again also observe:—

The stars have vanished, like the good done to a bad person; and the moonless quarters do not look charming, like women separated from their husbands. I think the sky itself, being intensely heated within by the fire of the weapon[23] of Indra, the lord of gods, is liquefied, and so is falling down in the form of water.

And also mark—

The cloud now rises, now descends, now pours, now thunders, and now produces a mass of darkness; thus it puts on manifold appearances, like a man getting wealth for the first time.

VITA: So it is.—

The sky is ablaze, as it were, on account of the lightning-flashes; it

22. Collyrium, in the archaic sense, is medicated eyewash. In these lines, the world is fancied as sleeping in the house formed of rain-showers under the canopy of clouds, with its eyes in the form of the beds of lotuses closed and its face in the shape of the circle of quarters muffled, with all distinction of night and day lost.

23. The thunderbolt.

laughs aloud, as it were, owing to hundreds of cranes; it is dancing, as it were, on account of Indra's bow, discharging arrows of water-torrents; it is shrieking, as it were, owing to the loud clap of thunder; it is reeling, as it were, because of the gales; and it is emitting (curling) smoke, as it were, on account of the throng of dark and snake-like clouds.

VASANTASENA:—

You are shameless, O cloud, since you frighten me, who am going to the house of my beloved, with your thunder, and touch me with your hands in the form of streams.

O Indra!—

Was I ever before attached to you amorously, that you are thundering at me with the lion-like roars of clouds? It is not proper that you should block my path with downpours of rain, when I am yearning for my beloved.

Moreover—

Just as, O Indra, you, being love-smitten, falsely said that you were Gautama, for the sake of Ahalya, so I too am suffering the pangs of love; therefore, O you who are indifferent to my pain, please draw off your obstructing clouds.[24]

Moreover—

You may thunder, O Indra, or pour down showers, or send down hundreds of thunderbolts: you cannot stop women who have started to meet their lovers.—

If the cloud thunders, well, let him do so; for men are known to be cruel. But, O Lightning, how is it that even you, who are a woman, cannot understand the sufferings of young women?

VITA: Madam, enough, enough of upbraiding her; she has laid you under an obligation:—

24. The allusion here is to Ahalya's seduction by Indra. Ahalya, the daughter of Brahma, was the wife of the sage Gautama. Indra once saw her and was completely enamoured by her beauty. On one occasion, as Gautama went out early in the morning for bathing and performing the proper religious duties, Indra deceived her by the device of assuming her husband's shape and ravished her.

Charudutta compares his longing for Vasantasena with Indra's longing for Ahalya.

For, like a slippery gold-chain on the chest of the Airavata[25], or like a white flag planted on the top of a mountain, this lightning, which is as it were the lamp illuminating the interior of Indra's abode, points out to you the residence of your lover.

VASANTASENA: Sir, it is as you say, here is that same house.

VITA: Since you have mastered all the arts, there is nothing on which you need advice in this matter. Still, my affection prompts me to speak. When you go in here, you must not show too much anger.—

If you get angry, there is no scope for love; but again, whence could there be charming love without anger? So you may get angry and anger him too; but at the same time you should be pacified and also pacify your lover.

Well, so much for that.

Who is there? Please inform the worthy Charudatta:—

'Here she, full of love and joy, and with her hair dripping with water, has come to the house of you, her lover, desirous of seeing you, at a time rendered charming by the clouds and fragrant with the *Kadamba* and the *Nipa* trees in bloom, terrified though she was by lightning and the clap of thunder; she is now busy washing her feet which are covered with mud that has stuck to her anklets.'

CHARUDATTA (*listening*): Friend, find out what this is.

VIDUSHAKA: As Your Honour commands. (*approaching* Vasantasena, *respectfully*). I greet you, madam!

VASANTASENA: Sir, I salute you! You are well seen! (*to* Vita) Sir, let this umbrella-bearer be for you only.

VITA (*to himself*): She has cleverly dismissed me by means of this plan! (*aloud*) All right. Lady Vasantasena.—

May you achieve success in easily disposing of your courtesy to your lover, which is the commodity offered for sale in the market of a courtesan's profession, which is the birthplace of hypocrisy, deceit, treachery and falsehood, which consists of perfidy, wherein love-sports have made their home, and where the stock-in-trade consists of the festival of union. (*exit*)

25. M. R. Kale mentions (dark) Airavata though generally it is a reference to Indra's white elephant.

VASANTASENA: Sir Maitreya, where is that gambler of yours?

VIDUSHAKA (*to himself*): Oh! Wonderful! By calling him a gambler, she has complimented my dear friend. (*aloud*) Lady, here he is, in the garden of dried-up trees.

VASANTASENA: Sir, what is that which you describe as your 'garden of dried-up trees'?

VIDUSHAKA: Lady, it is a place wherein one can neither eat nor drink!

(Vasantasena *smiles*)

VIDUSHAKA: So, you may enter it.

VASANTASENA (*aside, to her* Maid): What am I to say on entering it?

MAID: O gambler, do you find the evening delightful?

VASANTASENA: Shall I be able to say it?

MAID: The occasion itself will lend you strength.

VIDUSHAKA: Please enter, madam.

VASANTASENA (*entering and approaching, striking him with flowers*): O gambler, do you find the evening delightful?

CHARUDATTA (*seeing*): Ah! Vasantasena has come! (*joyfully getting up*) My darling—

My evenings always pass in wakefulness, and my nights too have always been passed in sighing. But today, O lady with elongated eyes, the evening has put an end to my grief, since I am now united with you!

So I welcome you—here's a seat, pray be seated here.

VIDUSHAKA: Here's a seat; Your Ladyship should sit down.

(Vasantasena *takes a seat, then all sit*)

CHARUDATTA: Friend, see, see!—

By the *Kadamba* flower which is hanging from the tip of her ear and is dripping rainwater, one of her breasts is besprinkled with water, as though it were a royal prince being crowned as heir-apparent.

Well, friend, Vasantasena's garments are wet; let another pair of excellent garments be brought for her.

VIDUSHAKA: As Your Honour commands.

MAID: Sir, Maitreya, you need not go. I shall wait upon Her

Ladyship myself. (*does accordingly*)

VIDUSHAKA (*aside, to* Charudatta): Friend, I want to ask Her Ladyship about something.

CHARUDATTA: Do so.

VIDUSHAKA (*aloud)*: Well, now, for what purpose has Your Ladyship come here in such darkness of a rainy day when the moonlight has vanished?

MAID: Lady, how simple is this Brahmana!

VASANTASENA: Say, rather, 'clever'!

MAID: This our lady has come to ask this: 'What is the price of that jewel-necklace?'

VIDUSHAKA (*aside, to* Charudatta): Sir, it was already said by me that the jewel-necklace was of little value, and the golden casket of greater value; she was not satisfied with that and has come to ask for something else.

MAID: For, it was lost by our mistress at gambling, believing that it was her own. And it cannot be discovered where that gaming-master, the king's emissary, has gone.

VIDUSHAKA: Madam, you are simply repeating what I myself said to you!

MAID: So, while we are looking for him, you may receive this golden casket in exchange. (*she shows it*)

(Vidushaka *remains in deep thought*)

MAID: Your Honour is gazing at it very fixedly. What then; was it ever seen by you before?

VIDUSHAKA: Madam, it captivates my eyes by its skilled craftsmanship.

MAID: Sir, your eyes are deceiving you! It is that very same golden casket!

VIDUSHAKA (*joyfully*): O friend, this is that very same golden casket that was stolen by the thieves from our house!

CHARUDATTA: Friend,—

The very device that was thought of by us for returning that deposit is now being used against us. But is it really true, or only a mockery?

VIDUSHAKA: O friend, it is true, — I swear by my Brahmanahood!

CHARUDATTA: It is gratifying, highly gratifying to me!

VIDUSHAKA (*aside, to* Charudatta): Sir, I'll ask her as to whence they obtained it.

CHARUDATTA: There is no harm in that.

VIDUSHAKA (*whispering in the* Maid's *ear*) Was it like that?

MAID (*whispering in* Vidushaka's *ear*): It was like that.

CHARUDATTA: What is this that is being communicated? Am I an outsider?

VIDUSHAKA (*whispering in* Charudatta's *ear*): It was like this.

CHARUDATTA: My good girl, is it really that very same golden casket?

MAID: Noble one, just so.

CHARUDATTA: Girl, I have never allowed a welcome announcement to go unrewarded. So, then, please accept this ring as a reward. (*Seeing his hand ringless, he exhibits shame*)

VASANTASENA (*to herself*): Hence it is that you are loved by me!

CHARUDATTA (*aside, to* Vidushaka): Alas, Oh!—

In this world, to a man who has lost his money, of what use is his very life in the first place?—Since he is unable to counter-act suitably, both his anger and his favour remain ineffectual.

Again—

A bird maimed of its wings; a dried-up tree; a lake without water; a cobra with its fangs pulled off; and a pauper: all these are on the same level, in this world.

And also—

Poor men are, indeed, like houses that are tenantless, wells devoid of water, and trees that are shrivelled, since, as they forget their penniless condition in the company of their beloved acquaintances, the moments of their gratification thus prove fruitless.

VIDUSHAKA: Sir, do not grieve too much. (*aloud, laughingly*) Madam, please return my bathing-garment.

VASANTASENA: Sir, Charudatta, was it indeed fair of you to test the worth of this person by that jewel-necklace?

CHARUDATTA (*with an embarrassed smile*): Mark, Vasantasena, mark!—

Who would have believed the real fact? Everyone would have thought lightly of me. For, in this world, inglorious poverty is always prone to be suspected.

VIDUSHAKA: Girl, are you going to sleep in this very place?

MAID (*laughing*): Sir Maitreya, you are now showing yourself exceedingly simple, indeed!

VIDUSHAKA: O friend, here now the rain has begun to fall again, in torrential showers of water, as if wishing to drive away people comfortably seated.

CHARUDATTA: You have observed truly.—

For, here are falling these showers of water, breaking through the cloud-spaces, like the spires of lotus-stalks piercing through the intervening mud, as though they were the showers of tears shed by the sky owing to the calamity that has befallen the Moon.

And again—

The clouds, looking dark like the garments of Balarama, are pouring down in showers which are very pure like the mind of a noble person, and which, being fierce, are hard like Arjuna's shafts, as if they are pouring forth Indra's collection of pearls.

See, beloved, see—

Here this red lightning, desiring to be united with the clouds, has appeared of its own will and is embracing the sky, like a beloved mistress who, being in love, desires union with her lover at the advent of clouds and goes to him, of her own will, embracing him; the sky which, like a lover, is anointed with these clouds resembling the paint of crushed *tamala* leaves and which is fanned by the continual, fragrant and cool breezes of the evening.

(Vasantasena, *displaying the emotion of love, embraces* Charudatta.)

CHARUDATTA (*displaying the pleasure of her touch, and embracing her in return*):—

Roar still louder, O cloud! For, by your favour this body of mine, suffering the anguish of love, has become as though it were full of *Kadamba* flowers, with its hair risen on end and its passion inflamed by Vasantasena's touch.

VIDUSHAKA: You rainy day, son-of-a-slave, you are, indeed, an ungentlemanly fellow, since now you are frightening Her Ladyship with the lightning!

CHARUDATTA: Friend, you should not rebuke it.—

Let the storm, pouring showers continually, last for a hundred years if it pleases, and so let the lightning flash too; for owing to them I have been embraced by this beloved lady, otherwise inaccessible to persons like me.

And moreover, O friend,—

Blessed, indeed, are the lives of those who closely press, against their own bodies, the bodies, wet and cooled by rain-water, of their mistresses come to their house of their own accord.

Dear Vasantasena,—

The canopy, as it is torn, is with great difficulty supported on the pillars, as the ends of their close-bricked pedestals have been loosened. And this painted wall, owing to its *chunam* paint being cracked, is thoroughly drenched by the torrent of water.

(*looking upwards*) Ah! Here's a rainbow! See, my dear, see!—

The sky has here given a yawn as it were,[26] with its tongue in the form of the lightning, its uplifted long arms in the shape of Indra's rainbow, and an expansive chin in the form of the clouds.

Come, then; let us enter the inside. (*gets up and walks about*)—

Shrilly on the *tala*[27] leaves, rumbling on the branches, harshly on stones and fiercely on the waters; thus do the showers fall, resembling lutes played upon in musical concerts to the keeping of time.[28]

(*Exit all*)

Thus ends Act V, entitled

'A Stormy Day'

26. A man when yawning often raises his arm, and so the rainbow is aptly considered as the arm of the sky. The chin also is extended in the act of yawning.
27. A tree whose leaves were used as a writing surface in ancient India.
28. The lute is struck so as to produce high-pitched or low-pitched or gentle musical notes, according to the variations in the voice of the singer, and to the marking of musical intervals.

ACT VI

(*Enter* Maid)

MAID: How is it that Her Ladyship does not wake up yet? Well, I shall go in and rouse her. (*gesticulates walking about*)

(*Enter* Vasantasena, *asleep, with her body covered up*)

MAID (*observing*): Get up, madam, get up! It is daybreak!

VASANTASENA (*awakening*): Ha! How can it be morning when it is still night!

MAID: This is morning *to us*; but to Your Ladyship it is still night.

VASANTASENA: Girl, and where is that gamester of yours?

MAID: Madam, the noble Charudatta has gone to the old garden, *Pushpakarandaka*,[1] after giving orders to Vardhamanaka.

VASANTASENA: What orders did he give?

MAID: 'Get the chariot ready at night, so that Vasantasena might depart.'

VASANTASENA: Girl, where am I to go?

MAID: Madam, where Charudatta has gone.

VASANTASENA (*embracing* Maid): Girl, I could not take a good view of him at night. So today I'll observe him with my own eyes. Girl, have I here entered the *antahpura*?[2]

MAID: Not only the *antahpura*, but you have entered the hearts of all persons here, too.

VASANTASENA: Do the servants of Charudatta feel distressed?

MAID: They will feel distressed.

VASANTASENA: When?

MAID: When Your Ladyship will depart.

VASANTASENA: Then I shall be the first to be distressed! (*persuasively*) Girl, take this jewel-necklace and go and give it to my sister, the worthy Dhuta. And say to her in my name: 'I am a slave,

1. Literally, 'the basket of flowers'. *Pushpakarandaka* garden is the place in *Mrichchhakatika* where Charudatta is supposed to meet Vasantasena.
2. See Glossary.

of His Honour Charudatta, bought by his merits; and so I am of you also. Let, therefore, this pearl-necklace be worn as an ornament round the neck by you.'

MAID: Madam, but Charudatta would get angry with Her Ladyship!

VASANTASENA: Go; he won't be angry.

MAID (*takes it*): As Your Ladyship orders. (*goes out and again enters*) Madam, Her Ladyship Dhuta says: 'This is given to you by my husband as a mark of his favour. So it would not be right for me to take it. And please be it known to you that my husband himself is my best ornament.'

(*Enter* Radanika, *with* Charudatta's *Son* Rohasena)

RADANIKA: Come, dear, we shall play with this little cart.

ROHASENA (*crying*): O Radanika, what do I care for this small clay-cart? Give me that same gold-cart.

RADANIKA (*sighing dejectedly*): Child, whence can we handle gold now? When prosperity will come to your father again, then you will play with a gold-cart. (*to herself*) Let me in the meanwhile try to amuse him. I'll go into the presence of Her Ladyship, Vasantasena. (*approaching*) Madam, I salute you!

VASANTASENA: Welcome to you, O Radanika! Well, whose son is this? The moon-faced one gladdens my heart, even though his body is unadorned with ornaments.

RADANIKA: He is the son of the worthy Charudatta, Rohasena by name.

VASANTASENA (*stretching forth her arms*): Come, my child, embrace me! (*seating him on her lap*). He resembles his father in his beauty.

RADANIKA: Not only in beauty, but I think in character, too. The noble Charudatta finds recreation in his company.

VASANTASENA: And for what reason is he crying?

RADANIKA: He was playing with a small gold-cart belonging to the son of a gentleman who lives next door. And that was taken away by him. Then, as he asked for it again, I made this clay-cart and gave it to him. Thereupon he says: 'O Radanika, what do I care for this small clay-cart! Give me that same gold-cart.'

VASANTASENA: Oh, alas! Oh, alas! He, too, should be distressed at the sight of other people's wealth! O divine Fate, you do indeed sport with the fortunes of men, resembling the drops of water fallen on lotus-leaves! (*tearfully*) Child, do not cry; you will play with a cart of gold.

ROHASENA: O Radanika, who is this lady?

VASANTASENA: A slave of your father, bought by his merits.

RADANIKA: Child, Her Ladyship is your mother.

ROHASENA: Radanika, you are telling a lie; for if Her Ladyship is my mother, then why has she put on ornaments?

VASANTASENA: Child, with your innocent lips you are saying something most touching. (*gesticulates taking off the ornaments; weeping*). Now here, I have become your mother! So please take these ornaments, and get a gold-cart made out of them.

ROHASENA: Get away! I won't take them! You are weeping!

VASANTASENA (*wiping her tears*): Child, I will not cry. Go and play. (*fills the clay-cart with her ornaments*) Child, get a golden cart made out of these.

(*Exit* Radanika, *with* Rohasena)

CHETA (*entering, mounted on a chariot*): Radanika, O Radanika! Please tell Her Ladyship Vasantasena, that the covered car is ready at the side-door.

(*Entering*)

RADANIKA: Madam, here this Vardhamanaka announces that the carriage is ready at the side-door.

VASANTASENA: Girl, let it wait for some time; in the meanwhile I shall get through my toilet.

RADANIKA (*going out*): Vardhamanaka, wait awhile, until Her Ladyship has finished her toilet.

CHETA: Ah, yes! I too have forgotten the carriage cushions. So in the meanwhile, I'll just go and come back with them. But these bullocks are restive because of their new nose-rope. Well, I'll go and come back in the car itself.

(*Exit* Cheta)

VASANTASENA: Bring me my toilet things, girl. I'll get through my toilet. (*remains engaged in dressing*)

STHAVARAKA CHETA (*entering, mounted on a cart*): Samsthanaka, the king's brother-in-law, has ordered me: 'Sthavaraka, take the car and come quickly to the old garden, *Pushpakarandaka.*' Well, then, I'll just go there—Speed on, bullocks, speed on! (*driving on and observing*) How! The village carts have blocked the road! Now, what shall I do here? (*haughtily*) You there, get out, get out of the way! (*listening*) What do you ask—'Whose car is that'? This is the car of Samsthanaka, the king's brother-in-law! So get out of the way quickly! (*seeing*) How! Here's a person who, on seeing me, suddenly concealed himself and made off in another direction, like a gambler, running away from the game at the sight of the gambling-master! So who may he be? Or, what have I to do with him? I'll go on quickly. You there, yokels, get out, get out of my way! What do you say—'Stop awhile, and give a turn to this wheel'? Eh, you! Am I, the brave servant of Samsthanaka, the king's brother-in-law, to help you to a twist of the wheel? Or rather, he is alone, poor chap. So I'll manage thus: stop this car near the side-door of the garden of the noble Charudatta. (*stopping the cart accordingly*) Here I am coming! (*exit*)

MAID: Madam, a sound like that of the rims of wheels is heard. So the carriage must have come.

VASANTASENA: Maid, go; my mind is in a hurry; so tell me the way to the side-door.

MAID: Come this way, madam, come!

VASANTASENA (*walking about*): Girl, you may now retire.

MAID: As Your Ladyship orders. (*exit*)

VASANTASENA (*showing that her right eye throbs,*[3] *and mounting the cart*): Why, I wonder, does this right eye of mine throb? Or, rather the sight of Charudatta himself will dispel all evil omens!

STHAVARAKA CHETA (*entering*): I have cleared off these blocking

3. The throbbing of the right eye, or of any part on the right side of a woman, is looked upon as an evil omen, and is supposed to indicate impending bad luck; while the opposite is the case with regard to males. The sight of a pious man, the mention of favourite deities, the worship of the Brahmanas *et al* are supposed to remove the effect of evil omens.

carts. So then I'll proceed. (*shows that he mounts and drives; to himself*) The vehicle is heavy! Or rather, it appears heavy to me as I have been fatigued with turning that wheel. All right I'll proceed. Push on, bullocks, push on!

(*Offstage*)

O you gate-keepers! Keep a vigilant watch at your respective guard-posts! For now this cowherd's son, Aryaka, has broken through the prison by killing his jailor; he has cut his chains, and has given us the slip; he is escaping! So capture him, do capture him!

(*Enter with a toss of the curtain in great hurry,* Aryaka, *veiled, and with a chain still clinging to one foot; he walks about*)

CHETA (*to himself*): There has sprung up a great commotion in the city. So I'll go very quickly. (*exit*)

ARYAKA:—

Like an elephant that has broken loose from his tying-post, I am now wandering about, dragging the one remaining fetter of the chain clinging to the forepart of my foot, having escaped from the terrible great ocean of the calamity of death commonly called 'the king's prison'.

Oh! I was forsooth brought from the cowherd's hamlet by King Palaka, frightened by the prophecy of a seer, and was then kept bound with a chain in a secret death-dealing dungeon. And I have now escaped from that imprisonment by the favour of my dear friend, Sarvilaka. (*shedding tears*)—

If I do possess good luck, then what fault is it of mine, that he should have imprisoned me, as one does a wild elephant? And moreover, one cannot oppose the workings of Destiny. But still, the king is one to whom all pay homage; what hostility can there be against such a powerful person?

So then, where shall I, the unlucky one, go? (*seeing*) Here is the house of some good man, with its side-door open!—

This house is in ruins; and this big panel of its gate has no belt fixed to it, and is cracked at its joints. Surely this householder, whose fortunes resemble mine, must have been recently reduced to penury!

(*Offstage*)

Push on, bullocks, push on!

ARYAKA (*listening*): Ah! A carriage is coming in this very direction!—

Can it be the carriage of a picnic-party, not occupied by ill-natured people? Or perhaps it is a bride's carriage, arriving to take her away. Or by good luck, it is to be taken outside the city, intended for some high personage. As it is unattended, I fervently hope that it may be empty, sent unto me by my good luck!

VARDHAMANAKA CHETA (*entering with the chariot*): Fine! I have brought those carriage-cushions. O Radanika, tell Her Ladyship Vasantasena, that she may get into the carriage which stands ready, and proceed to the old garden *Pushpakarandaka*.

ARYAKA (*listening*): This carriage belongs to a courtesan; and it is to go outside the city. All right; I'll mount it. (*approaches gently*)

CHETA (*hearing*): How! These are anklets jingling! So Her Ladyship must have come. (*to* Aryaka, *presuming him to be* Vasantasena) Madam, these bullocks are restive on account of their new nose-rope. So you should mount it from behind.

(Aryaka *does accordingly*).

CHETA: The jingling of the anklets shaken by the movement of the feet has now ceased; and the carriage has become heavy. So I take it that Her Ladyship must have now mounted. So then I'll proceed. Speed on, bullocks, speed on (*drives on*)!

VIRAKA (*entering*): Ho, there ! You, Jaya, Jayamana, Chandanaka, Mangala, Pushpabhadra, and the rest!—

How is it that you are walking about at ease? That cowherd's son who was put in jail, he is escaping, having simultaneously broken the heart of the king and his chains!

Eh, you, stand at the eastern gate of the main street! And you, too, at the western; and you at the southern; and you at the northern! And this part of the rampart here, to the same I'll go and mount, and observe the place around in the company of Chandanaka. Come, Chandanaka, come! This way, here!

CHANDANAKA (*entering, in excitement*): Ho, there! Viraka, Visalya, Bhimangade, Dandakalaka, Dandasura, and the rest of you!—

Come confidently; strive quickly and make haste, so that Royalty will be unable to move to another family!

And moreover—

In gardens, in gambling saloons, on roads, in the city, in markets, and in stables—quickly search every person about whom you may have a suspicion!—

Oh, O Viraka! What, what do you want me to see? You may speak without hesitation. Who is he that is rescuing that cowherd's son, having cut asunder his fetters?—

Who has the Sun in the eighth mansion, or the Moon in the fourth, or the planet Venus in the sixth, or Mars, the son of the Earth, in the fifth?—

Say, who has Jupiter in the sixth mansion of his birth-horoscope, or Saturn, the son of the Sun, in the ninth?[4] Who is he that is rescuing that cowherd's son, while Chandanaka is alive?

VIRAKA: Soldier Chandanaka!—

I swear by your heart, O Chandanaka, that someone must be carrying off that cowherd's son; for he was found missing when the sun was just half risen.

CHETA: Push on, bullocks, push on!

CHANDANAKA (*seeing*): Eh, there! Just see; see!—

A closed carriage is being driven along the middle of the high road; so just inquire to whom it belongs, and where it is going.

VIRAKA (*seeing*): You, driver of the cart, just stop driving that carriage! To whom does it belong? And who is seated inside? And whither is it going?

CHETA: Well, this carriage belongs to the noble Charudatta. Her Ladyship Vasantasena is seated within. And she is being taken to the old garden *Pushpakarandaka*, for sporting with Charudatta.

VIRAKA (*approaching* Chandanaka): This carriage-driver says that the cart belongs to the noble Charudatta, that Vasantasena is inside, and that she is being taken to the old garden *Pushpakarandaka*.

CHANDANAKA: Then allow it to go.

4. Pertaining to horoscope.

VIRAKA: Without being inspected at all?

CHANDANAKA: Yes. Why not?

VIRAKA: Trusting whom?

CHANDANAKA: The noble Charudatta.

VIRAKA: Who is the noble Charudatta or Vasantasena, that it is to go uninspected?

CHANDANAKA: Ah! Don't you know the noble Charudatta, nor Vasantasena? If you don't know the noble Charudatta or Vasantasena, then you don't know even the moon in the sky, with its light!—

What person does not know that Charudatta, whose virtues are as charming as lotuses, whose character is as attractive as the moon, who relieves the misery of the affected, and who is a jewel, the essence of the four oceans?—

In this city only two persons deserve to be honoured, and are an ornament to it—Her Ladyship Vasantasena, and Charudatta, the store of righteousness.

VIRAKA: O Chandanaka!—

I know Charudatta, and I know Vasantasena, too, quite well. But, when the king's business is to be done, I don't know even my own father!

ARYAKA (*to himself*): This person, Viraka, must have been my past enemy, and this one, Chandanaka, my past friend, since—

Although appointed to discharge the same function, these two do not behave in the same manner, just like the two fires, the one at a wedding and the other at a funeral.

CHANDANAKA: You are an investigating officer in the army, trusted by the king. I have held these bullocks; you may therefore inspect.

VIRAKA: You, too, are an officer in the army, trusted by the king. So you may inspect yourself.

CHANDANAKA: If I inspect, would it be as good as being inspected by you!

VIRAKA: What is inspected by you is inspected by King Palaka himself.

CHANDANAKA: You there, raise up the yoke!

(Cheta *does so*)

ARYAKA (*to himself*): Would the guards find me out? Unlucky that I am, I am without weapons, too! Or rather—

I shall imitate Bhima; my arm shall be my weapon! It is better to die fighting, rather than in chains, as a prisoner. Or rather, this is no time for rash enterprise.

(Chandanaka *gesticulates mounting the carriage, and looks inside*)

ARYAKA: I seek your protection, Sir!

CHANDANAKA (*speaking in Sanskrit*): Safety to one who seeks protection!

ARYAKA:—

Whoever deserts a person that seeks his protection, he is forsaken by the goddess of success, by his friends, and by all his relatives; and be is always exposed to derision everywhere.

CHANDANAKA: How! The cowherd's son Aryaka has, like a bird frightened by a hawk, fallen into the hands of a fowler. (*reflecting*) On one hand, this innocent man, seeking my protection and riding in the noble Charudatta's carriage, is the friend of the worthy Sarvilaka, who once saved my life. And on the other hand there is my official duty. So, now, what is the right thing to do under these circumstances? Or rather, come what may, I have already promised him freedom from fear.—

If ruin befalls a man who, intent on obliging others, promises safety to a frightened person, why, let it befall; but all the same, in the world it would be regarded as a righteous deed.

(*getting down in fear*) I have seen him—(*leaving it half-said*)—no, Her Ladyship Vasantasena. And she says: 'Is it proper, is it becoming, that I should be insulted on the high road while going to meet the noble Charudatta?'

VIRAKA: Chandanaka, a suspicion has arisen in my mind about this.

CHANDANAKA: How is it that you entertain a suspicion?

VIRAKA:—

Your voice sounded hoarse with nervousness when you said 'I have seen him', and immediately again, 'Her Ladyship Vasantasena.'

So I can't believe you in this matter.

CHANDANAKA: Ah, why should you not believe me? We natives of the south speak indiscriminately. We are conversant with the various dialects of different countries, as spoken by the Malechchha[5] tribes, such as Khasa, Khatti, Khada, Khadaththa, Vida, Karnata, Karna, Pravarana, Dravida, Chola, China, Barbara, Khera, Khana, Mukha, Madhughata, and the others; and we speak as we please, saying either 'I have seen *him*', or 'I have seen *her*'.

VIRAKA: Still, let me also inspect; for such is the king's order; and I have the king's trust.

CHANDANAKA: What, then, have I become one who is *not* trusted?

VIRAKA: Still, there's the king's order.

CHANDANAKA (*to himself*): If it were to get reported that the worthy cowherd's son escaped by riding in the carriage of the noble Charudatta, then our noble Charudatta would be punished by the king. So, what is the remedy in this matter? (*pondering*) I shall stir up a quarrel after the manner of the people in Karnataka. (*aloud*) You, Viraka, are you going to re-inspect what has been inspected by *me*, Chandanaka? Who are you, eh?

VIRAKA: Ho! And who are *you*?

CHANDANAKA: Honoured and respected as you are, you don't now remember your lineage.

VIRAKA (*angrily*): Ho! And what do you think is my lineage?

CHANDANAKA: Who can say?

VIRAKA: Speak.

CHANDANAKA: I'd rather not speak of it.—

Although I know what your lineage is, I do not speak it out, owing to the loftiness of my character. Let it therefore remain in my mind. What is the good of breaking open a *Kapittha* fruit?

VIRAKA: Nay, speak; out with it!

(Chandanaka *makes signs indicative of his caste*)

5. Here, Malechchha means a non-Aryan, barbarian tribe. Some of the tribes mentioned here, which must have been well known in the days of the poet, cannot be identified now but mostly are natives of South India.

VIRAKA: Ho! What is that?

CHANDANAKA:—

Once you held in your hand a broken piece of whetstone, trimming the scrubby beards of men, your fingers busy with the scissors. And now you have become an officer in the army!

VIRAKA: You, Chandanaka, you too, the honoured one, don't remember your own parentage.

CHANDANAKA: Ho! What is the parentage of *me*, Chandanaka, who am as pure as the moon?

VIRAKA: Who can say?

CHANDANAKA: Speak; out with it?

(Viraka *gesticulates and makes signs)*

CHANDANAKA: Ho! What is that?

VIRAKA: Listen, man, listen!—

Your parentage is most pure indeed! Your mother was a kettle-drum, your father was a drum, and you had a tambourine for your brother, O you ugly-faced one! And now you have become an officer in the army!

CHANDANAKA (*angrily*): Am I, Chandanaka, a shoemaker? Very well; inspect the carriage.

VIRAKA: You there, driver of the carriage, turn it back! I am going to inspect it.

(Cheta *does accordingly:* Viraka *is about to mount the* carriage: Chandanaka *suddenly seizes him by the hair, throws him down, and kicks him*)

VIRAKA (*getting up in rage*): You fellow, you have suddenly seized me by the hair and kicked me, while I, the trusted officer of the king, was executing the royal command. Ho! Just listen to what I say. If I don't see you drawn and quartered in the court of law, then I shall cease to be called Viraka!

CHANDANAKA: Oh, go to the king's palace or to the law-court! What do I care for you, who is like a dog?

VIRAKA: All right. (*exit*)

CHANDANAKA (*looking* in *all directions*): Go, go away, O carriage-driver! And, if any one inquires, you should say that the carriage was inspected and allowed to pass by Chandanaka and Viraka. And, madam Vasantasena, I'll give you this token. (*hands over his sword*)

ARYAKA (*takes the sword; joyfully, to himself*):

Oh! I have obtained a weapon. My right arm throbs. Everything is therefore going to be favourable to me. Oh! I am saved!

CHANDANAKA: Madam,—

I hope you'll remember Chandanaka, now that I have spoken to you and shown that I am to be trusted. I do not say this out of greed, but out of a feeling of affection.

ARYAKA:—

Chandanaka, possessing a noble nature like the moon, has today become my friend, by good luck. Well, Sir, I'll remember Chandanaka, if the prophecy of the seer ever comes true.

CHANDANAKA:—

May Siva, Vishnu, Brahma, the Sun and the Moon, all grant you safety, after killing all the followers of your foe, just as the goddess Parvati, after killing Sumbha and Nisumbha granted safety to the gods.[6]

(*Exit* Cheta *with the carriage*)

CHANDANAKA (*looking in the direction of* Cheta): Ah! As he left, my dear friend Sarvilaka too has gone, following just close at his back. Well, I have made an enemy of Viraka, the chief of the police, who is the king's confidant. So, then, I also will follow him, together with my sons and brothers. (*exit*)

Thus ends Act VI, entitled

'The Exchange of the Carriages'

6. Sumbha and Nisumbha were two brother demons who practised penance for thousands of years and propitiated Siva who conferred upon them a boon that in riches and strength they should excel the gods. In their exultation they waged a war against the gods and oppressed the world. The gods, at the advice of Brahma, Vishnu and Siva, appealed to Durga, who fought with them, defeated their forces, slew their commanders Chanda and Munda, and finally killed them. This story is given in the fifth and the following sections of the Chandipatha, of the *Markandeya-purana*.

ACT VII

(*Enter* Charudatta *and* Vidushaka)

VIDUSHAKA: Oh, just observe the beauty of this old *Pushpakarandaka* garden!

CHARUDATTA: Friend, it is as you say. For,—

The trees appear like merchants; the flowers stand out like the articles for sale; and the bees are wandering about, as if they were the men collecting the toll.

VIDUSHAKA: Sir, you may sit down on this stone-slab which is charming without any artificial decoration.

CHARUDATTA (*having seated himself*): Friend, Vardhamanaka tarries long!

VIDUSHAKA: Vardhamanaka was told by me to come very quickly with Vasantasena.

CHARUDATTA: Then why is he tardy?—

Can it be that another carriage is passing slowly in his front, and he is seeking room to pass it? Or, the axle being broken, is he replacing it by another? Or have the reins snapped? Or, is he waiting to have the road cleared, his path being blocked by logs of wood lying in the middle of the road? Or can it be that he is driving his two bullocks slowly, and is coming in a leisurely fashion as he pleases?

(*Entering with carriage, with* Aryaka *concealed therein*)

CHETA: Push on, bullocks, push on!

ARYAKA (*to himself*):—

Greatly frightened at the sight of the king's officers, and with my escape yet not fully affected owing to the chain still clinging to my foot, I am moving on, having mounted the good man's vehicle without the fact being known, and thus resembling a cuckoo reared by the female crows in their nest.

Oh! I have come far away from the city! So, then, shall I get down from this carriage and enter this dense grove of trees? Or shall I remain and see the master of the carriage? Or rather, let me not think of the dense grove of trees. It is reported that His Honour, the noble Charudatta is kind towards those who seek his help. So I'll leave after I have seen him personally.—

That good man will surely be gratified to see me, recently liberated from this calamity. It was owing to the virtues of that high-souled person that I saved this body of mine reduced to such a plight.

CHETA: This is that garden. So I'll go near. (*approaching*) Sir Maitreya!

VIDUSHAKA: Oh! I have good news for you! Vardhamanaka is calling. So Vasantasena must have arrived.

CHARUDATTA: It is welcome news to me, welcome, indeed!

VIDUSHAKA: You son-of-a slave, why are you so late?

CHETA: Sir, Maitreya, don't be angry; I was delayed as I had to go back and return, because of my having forgotten the carriage-cushions.

CHARUDATTA: Vardhamanaka, turn the carriage round. Friend Maitreya, you may help Vasantasena to get down.

VIDUSHAKA: Have her feet been tied by chains, that she cannot get down herself? (*getting up and opening the carriage*) Oh! This is not Mistress Vasantasena! Here's Mister Vasantasena!

CHARUDATTA: Friend, away with joking; for love brooks no delay. Or rather, I'll myself help her alight. (*he gets up*)

ARYAKA (*seeing*): Oh! Here is the owner of the carriage himself! He is charming not only to hear about, but also to look upon! Oh! I am saved!

CHARUDATTA (*mounting the carriage and seeing*): Oh! Who, then, may this person be?—

Having arms huge like the trunk of an elephant, shoulders plump and lofty like those of a lion, a chest very broad and even, and eyes reddish and restless; how did this high-souled person, of such noble appearance, come to such a state, not befitting his dignity, that he should be wearing a chain clinging to his foot?

Well, then, who may you be?

ARYAKA: I am Aryaka, a cowherd by birth, seeking your protection!

CHARUDATTA: What, the same that was brought by King Palaka from the hamlet of cowherds and imprisoned?

ARYAKA: None else.

CHARUDATTA:—

Fate itself brought you here; as you thus came within the range of my sight. I will abandon my life even, but not you who seek my protection.

(Aryaka *gesticulates delight*)

CHARUDATTA: Vardhamanaka, remove the chain from his foot.

CHETA: As Your Honour orders. (*does accordingly*) Sir, the fetters are removed.

ARYAKA: But other and firmer ones, made of love, have been put on!

VIDUSHAKA: Take up those fetters[1]. He has been set free; let us therefore now take our departure.

CHARUDATTA: Fie! Not so quick!

ARYAKA: Friend Charudatta, it was out of affection for you that I got in this chariot of yours; so kindly forgive it.

CHARUDATTA: I am honoured by you by thus voluntarily showing your love.

ARYAKA: I wish to go now, with your permission.

CHARUDATTA: You can go.

ARYAKA: All right; I'll get down.

CHARUDATTA: Friend, do not get down. As your fetters have been only just recently removed, you will not be able to walk with a quick gait. The carriage would also create confidence in this locality, where the officers are moving about in large numbers. So, you should travel in the carriage just as before.

ARYAKA: As Your Honour says.

CHARUDATTA: Go to your relatives in safety.

ARYAKA: Well, I have just acquired a relative in Your Honour.

CHARUDATTA: I should be remembered by you during the course of conversation.

1. It implies that Aryaka should move on with his task of usurping the corrupt King Palaka and Charudatta is in support of this.

ARYAKA: Can I forget my own self?

CHARUDATTA: May the gods guard you as you travel along the road!

ARYAKA: Rather it is you who have guarded me!

CHARUDATTA: You were guarded by your own good luck.

ARYAKA:—

Still, Sir, Your Honour was the cause even of that.

CHARUDATTA: Since there can be no great protection for a culprit, when Palaka is up and exerting himself, you should quickly make good your escape.

ARYAKA: So I'll go, hoping to meet again. (*exit*)

CHARUDATTA:—

It is undesirable to linger here even for a moment more, since we have thus committed a very grave offence against the king. O Maitreya, throw that chain in that old well; for kings can see such things through the eyes of their spies.

(*showing that his left eye throbs*[2]) Friend Maitreya, this person is eager to get a sight of Vasantasena.

Mark:—

As I haven't seen that beloved of mine today, my left eye is throbbing, and my heart feels uneasy, frightened without any apparent cause.

So come, let us go. (*walking about*) How! Just before me I see a Buddhist mendicant,[3] a sight of evil omen![4] (*reflecting*) Let him come in by this path; while we shall depart by this other path. (*exit*)

Thus ends Act VII, entitled

'The Escape of Aryaka'

2. Throbbing of the left eye portends evil in the case of men.
3. The sight of Sramanaka was considered inauspicious, probably because of his shaven and uncovered face. The Sramanaka is the same as Samvahaka of Act II. There are other historical reasons for referring to a 'Buddhist Mendicant' as 'a sight of evil omen'.
4. Avoiding a mendicant is in harmony with the Brahmanical doctrine; at the same time, it is clear that the period of absolute intolerance and persecution of Buddhists had not yet arrived.

ACT VIII

(*Enter a* Buddhist mendicant, *with a wet garment in his hand*)

MENDICANT: Ye, ignorant people, accumulate religious merit!—

Keep a check over your appetites; remain ever wide awake with the help of the drum in the form of meditation. For these cunning thieves in the form of the senses steal away religious merit stored for a long time.

Moreover, looking at the world through its aspect of transitoriness, I have taken refuge in religion.—

That man most surely attains heaven, who has killed five men, and, after killing a woman has guarded the town, and has also killed the helpless Chandala.[1]—

The head is shaved, the face is shaved, but the heart is not purified; then what is the good of shaving? But, if a person has his heart purified, then his shaved head becomes him very well, indeed!

I'll first enter this garden of the king's brother-in-law, and there in the lotus-pond wash this garment which has been dyed with red-coloured water; and then I'll very quickly come away. (*walks about and does accordingly*)

(*Offstage*)

SAKARA: Stop, stop, you rascal mendicant!

MENDICANT (*seeing, with fright*): Alack-a-day! Here comes that king's brother-in-law, Samsthanaka. Since a certain mendicant once happened to offend him, wherever now he sees any mendicant, the same or another, he bores a hole through his nose, as if he were a bullock, and then drives him away. So whom shall I approach for help, helpless that I am? Or rather, Lord Buddha himself will be my refuge!

(*Entering, with* Vita, *armed with a sword*)

1. Here, the five men are referred to as the five senses (sight, hearing, taste, smell and touch); a woman is referred to as ignorance (*avidya*) and Chandala as ego, and these three collectively lead the man astray. The doing away with *Avidya*, the purification of the mind, and the destruction of *Ahamkara* (ego), are necessary for the attainment of Moksha even according to the Vedanta theory.

SAKARA: Stop, stop, you rascal mendicant! I'll break your head, like that of a red radish[2] brought in the midst of a drinking party. (*he beats him*)

VITA: Bastard, it is not right to strike a mendicant who has put on red robes owing to his being weary of the world. So, why bother about him? Please glance at this garden, pleasant to visit.—

Noble deeds are done in it by the sylvan trees[3], which are both the refuge and the joy of homeless persons. It lies unguarded, like the heart of wicked persons; like a newly-acquired kingdom, it is one to be enjoyed without the restraint of conquest.

MENDICANT: Welcome! Be kind, O *Upasaka*[4]!

SAKARA: Mark, friend, mark; he is calling me names.

VITA: What does he say?

SAKARA: He calls me an *Upasaka*. Am I then a barber?

VITA: He is praising you by describing you as a worshipper of Buddha.

SAKARA: Praise, O mendicant, praise!

MENDICANT: You are *dhanya*; you are *punya*.[5]

SAKARA: Friend, he calls me *dhanya* and *punya*. Am I a materialist, or a watering trough, or a potter?

VITA: Bastard, he is forsooth praising you by saying that you are blessed and are holy.

SAKARA: Well, friend, why has he come here?

MENDICANT: To wash this garment.

SAKARA: Eh, you rascal mendicant, this *Pushpakarandaka* garden was given to me by my sister's husband, as being the best of all

2. It was a practice of drinkers to eat a red radish as a relish to induce thirst and add zest to liquor.
3. Relating to or characteristic of wooded regions. Also, a spirit that lives in or frequents the woods.
4. Worshipper of Buddha.
5. Sakara blunders on the words *dhanya* and *punya*. By *dhanya* he understands a materialist, an infidel, one who thinks of money only and by *punya*, a brick-trough for watering cattle.

gardens. Dogs and jackals drink water here. I, too, the best of mortals, and a man, do not bathe here. And here, in this lotus-pond, you are washing those horribly stinking garments, having a colour like the broth of old *Kulittha* grain.[6] So I'll kill you with one blow!

VITA: Bastard, I guess that he must have but recently turned an ascetic.

SAKARA: Friend, how do you know?

VITA: What is there to be known? See:—

Even now, having shorn his hair, the complexion of his forehead is bright-red as before. And, as only a short time has elapsed, on his shoulder no scar is produced by his ascetic robe. Moreover, he is not yet sufficiently practised in wearing his brown garments; and the end of his garment, which is very loose on account of the bagginess of the cloth, and which completely hides the middle part of his body, does not sit well on his shoulder.

MENDICANT: It is as you say, O Upasaka! I have but recently turned a monk.

SAKARA: Then why did you not become a monk the moment you were born? (*beats him*)

MENDICANT: My bow to Buddha!

VITA: What is the use of beating this poor fellow? Leave him; let him go.

SAKARA: Eh, just wait till I hold a consultation.

VITA: With whom?

SAKARA: With my heart.

VITA: Oh! He hasn't yet gone!

SAKARA: O my child, my heart, my master, my child! Should this mendicant be allowed to go, or should he be made to stay? (*to himself*) He should neither go nor stay. (*aloud*) Friend, I have consulted my heart. This is what my heart says—

VITA: What does it say?

SAKARA: He shall neither go nor stay! He shall neither breathe

6. A technical term in medicine which means the liquid turned viscous, obtained by boiling corn in water.

in nor breathe out! He shall immediately fall down here and die!

MENDICANT: My bow to Buddha! I seek your protection.

VITA: Let him go.

SAKARA: Yes, but on one condition.

VITA: What sort of condition?

SAKARA: He should throw mud in water in such a manner that the water does not turn muddy. Or else, he should collect the water in a heap and throw it in the mud.

VITA: What folly!—

This earth is bent under the weight of fools, whose thoughts and deeds are all perverse, whose bodies are simply like pieces of stones, and who are merely growths of flesh.

(Mendicant *gesticulates cursing*)

SAKARA: What is he saying?

VITA: He is praising you.

SAKARA: Praise, praise; do praise me once again.

(Mendicant *does accordingly and exit*)

VITA: Bastard, see the beauty of this garden:—

These trees, guarded by watchmen under the king's orders, decked with fruits and flowers, and encircled by full-grown, motionless creepers, are enjoying happiness, like men in the company of their wives.

SAKARA: My friend has spoken truly.—

The ground is picturesque owing to the numberless flowers fallen thereon; the trees are bent down with the load of blossoms; and the monkeys, hanging from the branches at the tops of the trees, look as graceful as *panasa*[7] fruits.

VITA: Bastard, here's a stone-slab; please sit down.

SAKARA: Here I seat myself. (*he sits down, and so does* Vita) Friend, even now I remember that Vasantasena. Like the words of a wicked man, she does not leave my mind.

VITA (*to himself*): Although he was spurned in that contemptuous manner, he still remembers her! Or rather:—

7. Jackfruit

In the case of low persons, their passion waxes stronger if they are spurned by women; but in the case of a good man, the same becomes milder, or even disappears altogether.

SAKARA: Friend, a long time has passed since I asked the Cheta, Sthavaraka, to come here very quickly with the carriage. He hasn't yet arrived. And I have been feeling hungry long since. It is now noon, and I cannot walk with my legs. Just mark, mark:—

The sun, occupying the middle of the sky, is difficult to gaze upon, like an enraged monkey. And the ground is extremely hot, like Gandhari[8], suffering great anguish when her hundred sons were all killed.

VITA: So it is.—

The herd of bullocks is sleeping under shade, letting fall their mouthfuls of grass. And the wild deer, suffering from thirst, are drinking the water of the lake, hot though it is. Men are not walking on the roads of the city, being exceedingly afraid of the heat. I think, therefore, that the carriage must have halted somewhere on its way, leaving off the heated track.

SAKARA: Friend,—

O friend, the rays of the sun have rested on my head; birds, feathered creatures, and sky-rangers are sheltering among the branches of the trees; and men, persons and people, heaving hot and deep sighs, are passing the period of the sun's heat by remaining in their houses and homes.

Friend, still! That Cheta hasn't come. I'll sing something to amuse myself. (*he sings*) Friend, friend, did you hear what I sang?

VITA: Why describe it? You are verily a Gandharva!

SAKARA: How should I not be a Gandharva?—

Cumin seed and nut-grass mixed with asafetida, and a piece of orris-root with ginger and treacle; this fragrant mixture I have taken;

8. She was the daughter of Subala, king of the Gandhara country, and wife of Dhritarashtra. As her husband was blind and lost the comforts of sight, she too wore a blindfold over her eyes to bring herself to his condition. A reference from the *Mahabharata*, wherein Gandhari's hundred sons were killed in the legendary war.

How, then, shall I not be sweet-voiced?

Friend, I'll sing once again. (*does so*) Friend, friend, did you hear what I sang?

VITA: What to say of it? You are verily a Gandharva!

SAKARA: How should I not be a Gandharva?—

I have eaten cukoo's flesh, seasoned with asafetida, with black-pepper powder added to it, fried in ghee, and besprinkled with oil and clarified butter; How, then, shall I not be sweet-voiced?

Friend, still the Cheta hasn't arrived!

VITA: You need not be anxious; he'll come just now.

(*Enter* Vasantasena *riding in the carriage, and* Cheta)

CHETA: I am rather afraid. The sun is in the middle of the sky; I hope the royal brother-in-law, Samsthanaka, won't be getting angry now! Let me, therefore, drive quickly—Push on, bullocks, push on!

VASANTASENA: Oh, alas! Oh, alas! This is surely not the voice of Vardhamanaka! What may it mean? Could it be that the worthy Charudatta, not wishing to weary the bullocks, has sent another driver and a different carriage? My right eye throbs, my heart quakes, and all the quarters look void. I see everything turned upside down!

SAKARA (*hearing the sound of the wheels*): Friend, friend, the carriage has arrived!

VITA: How do you know?

SAKARA: Don't you notice it, friend? It appears to be croaking, like an old hog.

VITA (*seeing*): You marked correctly. Here he comes!

SAKARA: Sthavaraka, my servant, my son, have you come?

CHETA: I have.

SAKARA: And the cart also has come?

CHETA: By all means.

SAKARA: And the bullocks, too, have come?

CHETA: Certainly.

SAKARA: And you, too, have arrived?

CHETA (*with a laugh*): Master, I, too, have arrived!

SAKARA: Then bring the carriage in.

CHETA: By what path?

SAKARA: Along this broken part of the wall.

CHETA: Master, the bullocks will die; the carriage will be smashed; and I, too, your servant, shall die.

SAKARA: Well, I'm the king's brother-in-law! If the bullocks die, I'll buy others; if the carriage is smashed, I'll get another built; and if you die, I'll find another carriage-driver.

CHETA: That would all be right; but then I shall cease to be my own self!

SAKARA: Eh, you, let everything perish; but bring in the carriage along the broken part of the wall.

CHETA: Break, your carriage, go to pieces along with the master! Let there be another carriage. I'll go to the master and inform him. (*entering*) How; It is not broken! Master, here comes your carriage!

SAKARA: Are not the bullocks snapped, and the reins dead, and you too have not perished?

CHETA: That's so.

SAKARA: Friend, come; we shall inspect the carriage. Friend, you are my preceptor, my most revered preceptor. I look upon you with respect, as one to be treated with confidence and honour. So you should get into the carriage before me.

VITA: So be it. (*he goes to mount*)

SAKARA: Or rather, you stand aside. Is it your father's carriage that you get in before me? *I* am the master of the carriage, and *I* therefore shall go in first.

VITA: But Your Honour said so.

SAKARA: Even though I did say so, still it would have been proper etiquette on your part to say: 'Master, please get in first yourself.'

VITA: Well, then, let Your Honour get in.

SAKARA: Here, now I'll mount it. Sthavaraka, my servant, my son, turn the cart around.

CHETA (*having turned it around*): May master mount it.

SAKARA (*mounts and looks in; shows fear, quickly dismounts, and clings to the neck of* Vita): Friend, friend, you're dead, you're dead! There is a demoness or a thief seated inside the cart. So, if she is a demoness, then we are stolen, both of us; and if it is a thief, then we are gobbled up, both of us!

VITA: Don't you fear. How can a demoness come to occupy this bullock-cart? May it not be that your eyes being dazzled with the heat of the midday sun, there arose this hallucination on your seeing the shadow of Sthavaraka, with the cloak on?

SAKARA: Sthavaraka, my servant, my son, are you alive?

CHETA: Yes, I am.

SAKARA: Friend, there is a female seated inside the carriage. You had better see it.

VITA: What! A woman?—

Let us pass away quickly by the road, with bent heads, like bullocks whose eyes are dazzled by the falling raindrops; for the eyes of me, to whom esteem in public assemblies is dear, are afraid to gaze upon high-born ladies.

VASANTASENA (*with surprise; to herself*): What! Here is the same royal brother-in-law, who is an eyesore to me! Then indeed I am in peril, I, the unlucky one! Now this visit of my unlucky self here has turned out fruitless, like a handful of seed sown in saline soil. So what shall I do here?

SAKARA: This old fellow, Cheta, is a coward; he won't peep in the carriage. Friend, you look into it.

VITA: Where's the harm? All right, so be it.

SAKARA: How! The jackals are flying about in the sky and the crows scampering along! So I'll run away before my friend is eaten up with eyes and gazed upon with teeth!

VITA (*seeing* Vasantasena; *sorrowfully, to himself*): Ah! How is it that the doe is following a tiger! Oh, alas;—

Leaving the male swan, resplendent like the autumnal moon, and reclining on a stretch of white sand, the female swan has taken herself to a crow!

(*aside; to* Vasantasena) O Vasantasena, this is neither proper nor becoming.

Having first proudly spurned him, you're now come at the bidding of your mother, for the sake of money—

VASANTASENA: No! (*she nods her head*)

VITA:—

Then I think you came owing to the profession of a courtesan being characterized by absence of generosity.

Why, I did say to you: 'O good maiden, serve equally one much coveted by you and one disliked by you.'

VASANTASENA: I came owing to an interchange of carriages. I beseech your protection!

VITA: Be not afraid; be not afraid: All right; I will deceive him. (*approaching* Sakara) Bastard, it is true; there's actually a demoness inside.

SAKARA: Friend, friend, if there's a demoness, then how has she not stolen you? And if a thief, then how haven't you been devoured?

VITA: What is the good of discussing that? Well, now, if we enter the city of Ujjayini going on foot, through the row of orchards, then what is the harm in that?

SAKARA: If we do so, what good would it do?

VITA: If we do so, we get exercise and avoid giving trouble to the bullocks.

SAKARA: So be it. Sthavaraka, my servant, take away the carriage. Or rather, stay—stay. Am I to walk on foot before the deities and the Brahmanas? No, no! I'll mount the carriage and go, so that, seeing me from afar, they will say: 'Here goes our master, the King's brother-in-law.'

VITA (*to himself*): It is difficult to convert poison into medicine. All right. Let it be thus. (*aloud*) Bastard, here's Vasantasena, come to you of her own accord, as an *abhisarika*[9].

VASANTASENA: Heaven forbid—heaven forbid!

SAKARA (*joyfully*): Friend, friend, has she come to *me*, the great man, a human being, another Krishna?

VITA: Yes.

9. See Glossary.

SAKARA: Then I have attained unprecedented good fortune! On that occasion I had angered her; now I'll fall at her feet and reconcile with her.

VITA: You have spoken well!

SAKARA: Here I'll fall at her feet. (*approaching* Vasantasena) Mother, mother, listen to my request:—

Here I prostrate myself at your feet, O you with large eyes, O pearl-toothed one, I place my folded hands on the ten nails of your feet. O beautiful-limbed damsel, I beseech your forgiveness for whatever ill I, tormented by love, did towards you. I am your slave!

VASANTASENA (*angrily*): Get away; you are speaking wickedly. (*strikes him with the foot*)

SAKARA (*wrathfully*):—

This head of mine, which was caressed by my mother, and which did not bow even to the gods, has been struck down with her foot, as a jackal treats a carcass in the jungle.

You, Sthavaraka, my slave, where did you come across her?

CHETA: Master the highway was blocked by the rustic carts. I therefore stopped my cart near Charudatta's garden; and while I got down there and was giving a turn to the wheels of another's cart, she must have got in by mistaking our carriage for her own; so I infer.

SAKARA: What! She came owing to an interchange of carriages, and not to meet me as her lover! Then get down, get down from my carriage! You are going to meet that penniless son of a merchant, and you make my bullocks carry you! Ah, get down, get down! You born servant-maid, get down!

VASANTASENA: To tell the truth, I feel honoured by these words that I am going to meet that worthy Charudatta. Now whatever must happen, let it come.

SAKARA:—

Seizing you by the hair with these two hands of mine, which are decked with a row of lotuses in the form of their ten nails, and which were yearning to pat you to the accompaniment of hundreds of fond words, I shall forthwith drag down your lovely body from my carriage, just as Jatayu[10] did the wife of Vali.

VITA:—

Ladies like these, who are accomplished, should not be seized by the hair; for, the creepers growing in orchards deserve not to have their foliage lopped off.

So you may stand off; I'll help her to dismount. O Vasantasena, please get down.

(Vasantasena *dismounts and stands in one corner*)

SAKARA (*to himself*): That fire of my anger, which was kindled on that occasion when she spurned my suit, has today been inflamed by this her insult of kicking. So I'll now kill her. All right; I'll proceed in this way. (*aloud*) Friend, friend!—

If you desire to wear a large cloak, with long fringes and having hundreds of tassels, and to dine upon flesh and derive satisfaction therefrom, munching with the sounds *chuhoo-chuhoo, chukku, chuhoo-chuhoo*.

VITA: Then what?

SAKARA: Do what I wish.

VITA: Very well. I'll do it, but excepting an execrable deed that cannot be done.

SAKARA: Friend, there is not even the remotest trace of the job being difficult, that cannot be done. There is no demoness whatsoever in it.

VITA: Speak out, then, what it is.

SAKARA: Kill Vasantasena!

VITA (*closing his ears*):—

If I kill her who is young, a woman, and an ornament of our city, and who, although a courtesan, bestows her love in a manner unlike that of the courtesans' quarter, and who is withal innocent, then with what boat shall I cross the river of the next world?

SAKARA: I will give you a boat. And moreover, in this deserted garden, who would see you killing her?

10. These references from *Ramayana* are incorrectly quoted by Sakara in his usual absurdity.

VITA:—

The ten quarters[11] are seeing me, and the sylvan deities, and the moon, and this sun with his brilliant rays, and Dharma, and Wind, and Sky, and the internal soul, and also the Earth, the witness to all holy and unholy acts.

SAKARA: Then put your cloak over her, and kill her.

VITA: Fool, you are damned!

SAKARA: This old jackal is afraid of committing sin. All right; I'll persuade my servant Sthavaraka. Sthavaraka, my slave, my son, I'll give you golden bracelets.

CHETA: And I will put them on.

SAKARA: I'll get a golden seat made for you.

CHETA: And I will sit on it.

SAKARA: I'll give you all the leavings of my food.

CHETA: And I will eat them.

SAKARA: I'll make you the head of all my servants.

CHETA: Master, I will become the head.

SAKARA: Then do my bidding,

CHETA: Master, I will do everything, excepting such as cannot be done.

SAKARA: There is no trace of impossibility about it.

CHETA: Let my master speak out what it is.

SAKARA: Kill this Vasantasena?

CHETA: Please, master, don't ask me that! This noble lady was brought here by my unworthy self owing to an interchange of carts.

SAKARA: You, slave, have I no mastery even over you?

CHETA: Your Honour is the master of my body, but not of my morals. So please, master, excuse me; I am verily afraid.

SAKARA: Being my servant, of whom are you afraid?

CHETA: Of the next world, master.

SAKARA: What is that 'next world'?

CHETA: Master, it is the consequence of one's good and bad acts.

11. Ten directions as according to Hindu mythology.

SAKARA: Of what kind is the consequence of good acts?

CHETA: To be like my master, adorned with many golden ornaments.

SAKARA: And of what sort is the consequence of bad acts?

CHETA: To be like me, who have become an eater of others' food. So I will not do an unholy deed.

SAKARA: Aye, you won't kill her? (*he beats him in a variety of ways*)

CHETA: The master may beat me or kill me; but I won't do an unholy deed.—

Since I am made a born slave by the faults of my luck, I shall not buy any further store of evil luck, and for this reason I refrain from committing a sinful deed.

VASANTASENA: Sir, I beseech your protection!

VITA: Bastard, forbear, forbear! Good, Sthavaraka, good!—

Even this penniless man, in such a miserable condition, and a menial, desires reward in the next world; but not so his master. Then why do such persons not meet with instant ruin, who multiply unworthy deeds and never do a worthy one?

Moreover—

Providence, ever searching for weak points is unjust, since slavery is Sthavaraka's lot and mastery yours: and since he does not enjoy your wealth and you are not made to do his bidding.

SAKARA (*to himself*): The old jackal is afraid of unrighteousness, and this born slave is afraid of the next world. But I am the king's brother-in-law, an aristocratic gentleman! Whom do I fear? (*aloud*) You born slave, Cheta, you go away. Getting into a secluded spot, you wait there, enjoying rest in solitude.

CHETA: As the master orders. (*approaching* Vasantasena) Lady, only thus far lay my power. (*exit*)

SAKARA (*girding up his loins*): Stay, Vasantasena, stay. I am going to kill you!

VITA: Ah, will you kill her in my presence? (*he seizes him by the neck*)

SAKARA (*falls on the ground*): My friend is killing me, the master.

(*shows that he faints; regaining consciousness*)—

I have always fattened him on meat and ghee; but now that a business has arisen, how is it that he has turned my enemy?

(*reflecting*) Well, I have found the remedy. The old jackal has given me a warning by shaking my head. So I'll first send him away and then kill Vasantasena. I'll proceed in this way. (*aloud*) Friend, as to what I said to you; how can I, born in such a very large family, of the size of a *mallaka*[12], ever do such a wicked deed? I spoke simply to force her into compliance with my proposal.

VITA:—

Why talk of one's family, since in such matters it is a man's character that makes him what he is? Even thorny trees grow exceedingly large in a fertile field.

SAKARA: Friend, she feels bashful in your presence, and therefore does not accept me. So you should take yourself off. The slave Sthavaraka, whom I beat, has already gone. See, here he goes away running. So please catch him and then come back with him.

VITA (*to himself*):—

Vasantasena would not accept this fool in my presence, out of a sense of pride. So, I'll just secure privacy for her by going away: for love becomes enjoyable only through the confidence born of privacy.

(*aloud*) All right. I'll go.

VASANTASENA (*seizing him by the hem of his cloak*): I repeat, sir, that I beseech your protection!

VITA: Fear not, fear not, O Vasantasena! Bastard, let Vasantasena remain a pledge in your hands.

SAKARA: Yes. Let her remain in my hands as a pledge.

VITA: Honestly?

SAKARA: Honestly.

VITA (*going a little distance*): Or rather, this wicked fellow might kill her when I am gone. So I'll conceal myself and just observe what he intends to do. (*he remains* in a corner)

SAKARA: Well, now I'll kill her. Or rather, this cunning rascal

12. It means a small vessel or cup formed of leaves or it may mean a wine cup.

Brahmana, this old jackal, might perhaps go and hide himself, and, behaving like a fox, might cheat me. So I'll just do this to deceive him. (*he collects some flowers and decorates himself*) Girl, girl, Vasantasena, come!

VITA: Oh! He has turned a gallant! Ah! Now I feel at ease. I'll go now. (*exit*)

SAKARA:—

I give you gold, speak caressingly, and fall down at your feet with my head having the turban on. Still, O damsel with pure teeth, why don't you desire to accept me as your servant? Men, alas, have to suffer anguish!

VASANTASENA: What doubt is there about that?

(*With face bent down she recites these two verses, beginning 'You man of evil deeds' et al*)—

You man of evil deeds, you wretch! Full of sin that you are, why do you tempt me with money, in this matter? For the bees never abandon a lotus of graceful actions and pure form.—

An effort should be made to serve a man of good family and virtuous character, even though he may be in reduced circumstances. For, love for a worthy personage constitutes the glory of courtesans.

Moreover, having resorted to the mango tree, I shall not betake myself to the *Palasa*[13] tree.

SAKARA: You slave-born maid, you've called that penniless Charudatta a mango tree, while me you have described as a *Palasa,* and not even as a *Kimsuka!* You, who have been thus calling me names, —still you remember that same Charudatta?

VASANTASENA: When he is enshrined just in my heart, why should he not be remembered?

SAKARA: Now I'll crush you, and him who is still stationed in your heart, both together! So wait, O you who love that penniless

13. Also called Kimsuka. The Palasa has bright red flowers which have no smell; so by comparing Sakara to the Palasa tree, Vasantasena meant to imply that he had wealth but no merits. But Palasa also meant a raw-flesh eater, a demon. So Sakara thought that she purposely used the word Palasa in order to indirectly call him a demon, and herein lay the abuse.

merchant-man, wait!

VASANTASENA: Speak, speak; speak out once more those charming syllables!

SAKARA: Now let that slave-born penniless Charudatta save you if he can!

VASANTASENA: He will save me if he will see me!

SAKARA:—

What! Is he Indra, the great Indra, the son of Vali; or Kalanemi, the son of Rambha; or Subadhu; or King Rudra; or Jatayu, the son of Drona; or Chanakya, or Dhundhumara, or Trisanku?[14]

Or why? Even these can't save you!—

Just as in the Bharata age, Sita was killed by Chanakya, I'll kill you, like Jatayu killed Draupadi!

(*he is about to strike her*)

VASANTASENA: O mother, where are you? O noble Charudatta, this person is now dying, with her desires still unfulfilled! Now I will cry out loudly for help. Or no, it would be surely a disgrace that Vasantasena should be crying aloud—bow to the noble Charudatta!

SAKARA: Still the born slave is calling out the name of that same villain! (*he seizes her by throat*) Remember, you born slave, remember!

VASANTASENA: Bow to the noble Charudatta!

SAKARA: Die, you born slave, die! (*gesticulates strangling her and killing her*)

(Vasantasena *faints* and *falls down senseless*)

SAKARA (*gleefully*):—

I have killed that wicked girl, who was a repository of faults and a dwelling place of immodesty, and who, full of love came to sport with that Charudatta, who had come here, her end being near. But why am I thus describing the valour of my arms? The poor woman dies even when deprived of her breath, so as never to rise again, as did Sita in the Bharata age.—

Because she wouldn't accept me who was courting her, I killed in my wrath that courtesan. Seeing that the *Pushpakarandaka* garden was

14. Sakara's incorrect references from Hindu mythology.

lonely, I suddenly strangled her with the noose of my arms. My brother and my father, and my mother like Draupadi, who have not witnessed this sort of valorous exploit of their son, have indeed, missed a treat to their sight.

Very well. Now that old jackal would come back; so I'll move away hence and wait in another spot. (*does accordingly*)

VITA (*entering, with* Cheta): I have persuaded Cheta Sthavaraka to come back. So I'll now see the bastard. (*walking about and observing*) Ha! Here's a tree fallen just in our way! And while falling it has killed a woman! You sinful tree, what a wicked deed you have committed! Now, we too have been completely brought down into ill luck, by the sight of the woman killed by the fall of your sinful self. This is a bad omen; so, to tell the truth, my mind is full of misgivings about the fate of Vasantasena. May the gods grant safety in all ways! (*approaching* Sakara) Bastard, here I have induced the Cheta Sthavaraka to come back.

SAKARA: You're welcome, friend. My son, my slave, Sthavaraka, you too are welcome!

CHETA: Thank you.

VITA: Return my pledge.

SAKARA: Pledge? Of what sort?

VITA: Vasantasena.

SAKARA: She's gone!

VITA: Where?

SAKARA: Just immediately after you.

VITA (*surmising*): Surely, she couldn't have gone in that direction!

SAKARA: In what direction did you go?

VITA: In the east.

SAKARA: Well, she went towards the south!

VITA: I went to the south!

SAKARA: She went to the north!

VITA: You are telling something that is extremely contradictory. My mind does not feel at ease. So tell me the truth.

SAKARA: I swear it by your head and by my feet. So compose

your heart. I killed her!

VITA (*sorrowfully*): You have really killed her?

SAKARA: If you don't believe in my words, then observe this first exploit of Samsthanaka, the king's brother-in-law! (*shows her body*)

VITA: Alas, I am undone, unlucky that I am! (*swoons and falls down*)

SAKARA: Ho, ho! My friend has breathed his last!

CHETA: Compose yourself, sir, compose yourself! It was really I who first killed her by thoughtlessly bringing that carriage here.

VITA (*coming to himself; mournfully*): O Vasantasena!—

The river of the water in the form of innocence is dried up. Love has departed to its original home! Alas, alas! O you who were an adornment of your ornaments, you with that sweet face, you that shone with your fondness for amorous sports! You that were the river of goodness, with laughter for the sandy bank, the refuge of persons like me! Alas, alas! The market of passion, the treasure-house of merchandise of beauty, is now pulled down!

(*tearfully*) Alas, Oh, alas!—

What indeed could be the purpose for which you committed this deed? The sinless beauty of this city has been felled by you who are almost an embodiment of sin.

(*to himself*) Ah, this wicked fellow might perhaps shift on me the guilt of this criminal deed. All right; I'll leave this place. (*walks about*)

(Sakara *approaches and holds him*)

VITA: You sinner, don't touch me! I have had enough of you! Let me go!

SAKARA: Ah you! Having, yourself killed Vasantasena, where do you escape now, putting the guilt on me? Now I have been rendered so very helpless.

VITA: Oh, go to the devil!

SAKARA:—

I will give you money to the extent of a hundred gold coins; I will give you a *karshapana*[15], together with a *vodi*[16]. Let this, my exploit,

15-16. Names of different coins.

which is the abode of crime, be common to all persons.

VITA: Fie! Let it belong to you alone.

CHETA: Heaven forbid!

(Sakara *laughs*)

VITA:—

Let there be absence of joy; give up this laughter; fie upon your friendship which is humiliating and dishonourable! May I never come in contact with you again! I cast you off, who are without *guna*[17], like a broken bow[18].

SAKARA: Friend, please be not angry. Come; we'll descend into this lotus-pool and sport in its waters.

VITA:—

The people of this city consider me as if degraded and wicked while serving you even when you had not fallen so low as now. How, then, can I still follow you, who have now killed a woman, and who will again be fearfully looked at with half-closed eyes by the women of this city?

(*sorrowfully*) Vasantasena!—

O fair one, may you never become a courtesan in your next birth; O you possessed of the excellences of character, may you be born in a pure family!

SAKARA: Where are you now escaping, having killed Vasantasena in my old *Pushpakarandaka* garden? Come, you must make your deposition before my brother-in-law the king! (*seizes him*)

VITA: Ah, just wait, you villain! (*draws his sword*)

SAKARA (*moving off in fear*): What, are you afraid? Well, you may go.

VITA (*to himself*): It is not desirable to stay here any longer. All right; I will go to the place where there are the worthy Sarvilaka, Chandanaka, and others. (*exit*)

SAKARA: Go to perdition! Well, Sthavaraka my son, what sort of a deed have I done?

17. Here, void of virtues.

18. A broken bow is without string and is of no use, hence virtueless.

CHETA: Master, you have committed a most heinous crime!

SAKARA: What, you slave, what do you say? That I committed a crime? All right; I will manage thus. (*takes off many ornaments*) Take these ornaments, which I hand over to you. They are mine during the time that I have to decorate myself; on other occasions they shall be yours.

CHETA: These look well on my master's own body; of what use are they to me?

SAKARA: Well, then, be off; take away those bullocks and stay in the aisle of the elephant-corniced upper terrace of my palace, until I come there.

CHETA: As my master orders. (*exit*)

SAKARA: My friend has made himself scarce, to save his own skin. Now this slave also I am going to put in fetters and keep imprisoned in the elephant-corniced terrace room of my palace. Thus my secret will be kept safe. Now let me go; or no—first let me have a look at her, whether she is dead or whether I shall have to kill her again. (*observing*) How! She is quite dead! Well, now, I'll cover her up with this cloak. But, no,—this is marked with my name: so some gentleman might recognize it. All right; I'll cover her up with this lot of dried leaves which have been heaped together by the gusts of wind. (*does so; reflecting*), Well I'll now proceed in this way: I shall forthwith go to the law-court and file a written charge that the merchant Charudatta enticed Vasantasena into my old garden, *Pushpakarandaka*, and killed her for the sake of her money.—

I'll invent a new fraud, for the destruction of Charudatta, horrible like the slaughter of a cow in a holy city.

All right; I'll depart. (*goes out and sees; with fear*) Alack-a-day! By whatever path I try to leave, by that same path I find this rascal Buddhist mendicant coming in, with his garment dyed with red-coloured water. He is inimical towards me, as I had bored a hole through his nose and turned him out; seeing me, perhaps he might give out that it was I who killed her. So how shall I leave? (*seeing*) Yes. I'll jump over this half-fallen broken wall, and take myself off.—

Here I go with very great speed, like monkey Mahendra going to the city of Lanka, through the sky, over the earth, through the nether regions, and over the summit of the mountain Hanumat. (*exit*)

(*Entering, with a toss of the curtain,* Samvahaka, *now appearing as a Buddhist* Mendicant)

MENDICANT: I have washed this piece of cloth. What, shall I dry it by hanging it on the branch of some tree? Here the monkeys might make off with it. Shall I then spread it over the ground? There it would be spoiled by the dust. So, where shall I spread it for drying? (*seeing*) All right; I'll spread it here on this heap of dry leaves collected together by the gusts of wind. (*does so*) Bow to Buddha! (sits *down*) Well, now, let me recite holy words. (*he recites the verses beginning with 'That man most surely attains heaven' et al, which he had repeated before*) Or no—I have had enough of this seeking after heaven; so long as I do not requite the obligations of that Vasantasena, the devotee of Buddha, who bought me off from those two gamesters by paying ten gold-pieces. From that time I consider myself as having been purchased by her as a slave. (*seeing*) Ha! What may this be, that is heaving inside these leaves? Or rather—

I think that these leaves are throbbing as they were first heated by the hot breezes and then moistened by the water of my garment, the leaves which resemble wings of birds with their feathers extended.

(Vasantasena *regains consciousness and puts forth her hand*)

MENDICANT: Ha! What! A woman's hand is coming out, adorned with fine ornaments! What! Another hand, too! (*observes it variously*) I seem to recognize this hand. Or why, what doubt can there be? Surely it is that same hand which granted me safety! Well, I will see. (*gesticulates uncovering it; observing and recognizing*) It is she, that same devotee of Buddha!

(Vasantasena *shows a desire for water*)

MENDICANT: How! She asks for water! And the well is far off! What shall I now do about this? All right; I will let this garment drip over her. (*does so*)

(Vasantasena *regains consciousness and rises.* Mendicant *fans her with the hem of his cloak*)

VASANTASENA: Sir, who may you be?

MENDICANT: What, does not Your Ladyship, the worshipper of

Buddha, remember me whom you purchased with ten gold-pieces?

VASANTASENA: I do remember now. But it isn't as you say: It were far better that I was dead!

MENDICANT: O worshipper of Buddha, what may this be?

VASANTASENA (*with dejection*): What befits the profession of a courtesan.

MENDICANT: Please stand up, O worshipper of Buddha, stand up, by supporting yourself with this creeper growing near this tree. (*bends down the creeper*)

(Vasantasena *seizes it and stands up*)

MENDICANT: In this *Vihara*[19] near by, there dwells a nun, sister of mine by faith. You, the devotee of Buddha, should recover the composure of your mind in that place, and then go home. So, please walk very slowly, O worshipper of Buddha! (*walks about; seeing*) Make way, worthy sirs, make way! Here's a young lady, and here's a mendicant; still this is my righteous duty.—

He is really to be called a man who is restrained in the use of his hands, his mouth, and his senses. What can the king's law-court do to him? The next world lies firmly grasped in his hand.

(Exit All)

Thus ends Act VIII, entitled

'The Killing of Vasantasena'

19. Buddhist monastery.

ACT IX

(*Enter* Sodhanaka, *the beadle*)

SODHANAKA: The officers of the law-court have ordered me: 'O Sodhanaka, go to the judicial hall and see that the seats are ready.' So then I will just go to the judicial chamber to make them ready. (*walking about and seeing*) Here is the court of justice. I will just enter it. (*enters, sweeps it and arranges the seats*) I have swept the court-hall clean and arranged the seats. So I will now go and report the same to the judges. (*walks about; seeing*) How now! Here's that wicked fellow, villain and rogue, the king's brother-in-law, coming in this very direction! So I will first keep out of the range of his sight and then go. (*he stands aside in a corner*)

(*Enter* Sakara, *dressed in gaudy attire*)

SAKARA:—

I have bathed with water, fluids and liquids, and sat in orchards, gardens and groves in the company of ladies, damsels and women, like a Gandharva having well-shaped limbs.[1]—

For one moment I have my hair knotted; in the next I wear it matted and bound up like an ascetic. For one moment I have long, flowing tresses; in another curly hair. One moment I keep them loose; in the next I have them rolled up on my crest. Thus I am a picturesque and a wonderful brother-in-law of the king!

Moreover, I have found a broad way out, like a worm seeking a way, having got within the interior of the knotted root of a poisonous herb. So on whom shall I shift this evil deed of mine? (*recollecting*) Yes, I remember it! I'll lay the blame of this vile act on the penniless Charudatta! Moreover, he is known to be poor; everything will be believed as possible in his case.—All right. I'll go to the court of justice and lodge a written complaint, before anyone else has done so, that Charudatta strangled Vasantasena and murdered her. So, then, I'll just return to the law-court. (*walking about and observing*) Here's that law-court! I'll enter it. (*entering and looking on*) How! They have kept

1. Gandharvas are known for singing and scarcely for the beauty of their limbs; the usage here is attributed to Sakara's ignorance.

the seats ready arranged! So, until the judges arrive, I'll sit down here for a while in this grassy quadrangle, and wait. (*stays accordingly*)

SODHANAKA (*walking in another direction; looking before him*): Here the judges are coming! So then I will approach them. (*draws near*)

(*Enter a* Judge, *accompanied by a* Sreshthin,
a Kayastha, *and others*)

JUDGE: Sirs, Sreshthin and Kayastha!

SRESHTHIN & KAYASTHA: We await Your Honour's commands!

JUDGE: See, it is verily difficult for judges to find what lies in the minds of others, as a judicial trial is dependent on a number of other matters.—

The contending parties bring up for trial transactions whose real character is hidden and far removed from justice. Themselves swayed by self-interest, they never tell in the court their own faults. So the king incurs the blame for the failure of justice, whose intensity is augmented by the contentions of one party as well as the other. In short, nothing but odium is commonly to be got by a judge, appreciation but seldom falling to his lot.

Moreover—

People, completely ignoring law, wrathfully complain of the offences of others, but they never confess in the court their own faults. Even learned persons counsels who commit sin, sharing in the crimes of the contending parties, are verily lost. In short, blame is very easy for a judge to get; applause is far removed from him, seldom falling to his lot.

Hence, a judge—

Should be learned in law; expert in tracing the course of fraud; eloquent; never losing his temper; equally impartial towards friends, strangers and relatives; giving his decision only after carefully investigating the facts; a protector of the weak; a terror to the rogues: righteous; free from greed even when the means exist for him to take bribes; sincerely bent in his mind on discovering the real truth, and able to avert the anger of the king.

SRESHTHIN & KAYASTHA: Can they speak of merits as faults, even in Your Honour's case? If so, then they would say that there is darkness even when there is moonlight.

JUDGE: Sodhanaka, my good man, point out the way to the court of justice.

SODHANAKA: Please come, come; step this way, Your Worship!

(*They walk about*)

SODHANAKA: Here is the court of justice. So may the officers enter.

(*All enter*)

JUDGE: Sodhanaka, my good man, just go out and ascertain who there are that desire to present cases today.

SODHANAKA: As your worship orders. (*going out*) Gentlemen, the judges ask, who are here that desire to present cases?

SAKARA (*joyfully*): The judges have arrived (*walking forth arrogantly*): Here am I, the best of men, a gentleman, another Krishna, the king's brother-in-law, the brother-in-law of the king, who want to present a case!

SODHANAKA (*in flurry*): Ah! At the very commencement here's the royal brother-in-law desiring to present a case! All right. Sir, please wait awhile, so that I will just announce your presence to the judges. (*going to them*) Gentlemen, here is this king's brother-in-law come to the court, who wishes to present a case.

JUDGE: How! At the very commencement there is the king's brother-in-law wanting to present a case! It is just like an eclipse at sunrise, foretelling the death of some great man! Sodhanaka, the hearing of cases today will prove troublesome! Go out, my good man, and say to him that he might go away, as his suit can't be heard today.

SODHANAKA: As Your Honour orders. (*going out and approaching* Sakara) Sir, the judges say that you might go away as your suit won't be heard today.

SAKARA (*wrathfully*): What, ho! My suit won't be heard? If it be not heard, then I will inform my brother-in-law, the King Palaka, who is my sister's husband, and I will tell my sister and my mother, and I'll have this judge removed and another appointed in his place. (*he is about to leave*)

SODHANAKA: Sir, royal brother-in-law, stay awhile, so that I will just inform the judges of what you have said. (*approaching the* Judge) Our royal brother-in-law here says in anger—(*repeats what was said by him*).

JUDGE: Everything is possible in the case of that fool. My good man, say to him that he might come in, and that we will look into his suit.

SODHANAKA (*approaching* Sakara): Sir, the judges say that you might come in, and that they will look into your suit. Please, sir, you may therefore go in.

SAKARA: First they say that they wouldn't look into it, and now that they would. So, obviously, the judges must have got mightily frightened! Whatever I'll say, the same I'll compel them to accept as true. Very well; I'll go in. (*entering and approaching*) I am doing very well! As to you, I can give or not give you happiness as I please!

JUDGE (*to himself*): Oh! Even when presenting a complaint, his usual habits do not leave him! (*aloud*) Pray seat yourself.

SAKARA: Ah! This place belongs to me; so I'll sit where I like! (*to the* Sreshthin) I'll sit here! (*to* Sodhanaka) No, I'll sit here! (*then placing his hand on the head of the* Judge) I'll sit here! (*thereupon he sits on the ground*)

JUDGE: So you wish to lodge a complaint?

SAKARA: Yes.

JUDGE: Then tell us what it is.

SAKARA: I will whisper my business in your ear. I am born in a very big family, of the size of a *mallaka.*—

My father is the king's father-in-law; the king is the son-in-law of my father. I myself am the king's brother-in-law; while the king is the husband of my sister.

JUDGE: All that is known to us—

Why describe one's family, since one's own character itself is the determining factor herein? Even thorny trees grow exceedingly large in a good soil.

So describe your business.

SAKARA: I will just tell you. But understand that the king won't do anything to me even if I am guilty! Well, that brother-in-law of mine, being pleased with me, bestowed upon me the old garden *Pushpakarandaka*, the best of all gardens, for sport as well as for custody. And I daily go there, to look after it, to see that it is properly drained, and swept, and growing luxuriantly, and trimmed as necessary. As fate would have it, I saw therein—or no, I didn't see—the body of a woman lying dead on the ground.

JUDGE: And could you discover who that dead woman was?

SAKARA: O you judges, how could it be that I shouldn't recognize that lady, so well-known, the adornment of our city and wearing hundreds of ornaments? Vasantasena, who was enticed into the old *Pushpakarandaka* garden, when it was deserted, was killed by some villainous fellow for the sake of filthy lucre after having forcibly strangled her with the noose of his arms; but not by me—(*leaving it half-said, he covers his mouth*)

JUDGE: What negligence on the part of the city's police! O Sreshthin and Kayastha, first write down the words '*Not by me*' as part of the plaint.

KAYASTHA: As your worship orders. (*does so*) Sir, I have written it.

SAKARA (*to himself*): Hang it all! I have today simply brought ruin on my own self, like the beggar gulping down the hot *payasa*[2] in a hurry. Nevermind; I will proceed in this way. (*aloud*) O judges, what I was saying was that it was by me that she was seen. Why do you make such a fuss about it? (*with his foot he rubs out what was written*)

JUDGE: How do you know that she was killed for her money, and with the noose of arms?

SAKARA: Well, surely, I could guess it from her neck, which was very much swollen, and bare in the place where an ornament is worn, and also from her limbs where ornaments are worn, but which were devoid of golden ornaments.

SRESHTHIN & KAYASTHA: It seems quite plausible.

SAKARA (*to himself*): Most fortunate! I feel I have obtained a new

2. *Payasa* is a dainty preparation of rice boiled in milk with sugar. The story of a beggar who in his voracity swallowed hot milk-dish and killed himself, must have been popular in Sudraka's time.

lease of life! Heigh ho!

SRESHTHIN & KAYASTHA: Sir, on whom does the investigation of this suit depend?

JUDGE: Well, in law the complaints are of two kinds.

SRESHTHIN & KAYASTHA: What are they?

JUDGE: One depending on arguments and the other on actual facts. The one which depends on argument is to be settled by the plaintiff and the defendant. And what depends on facts is to be decided by the Judge with the help of his own wisdom.

SRESHTHIN & KAYASTHA: So, then, this case depends on the mother of Vasantasena.

JUDGE: That's so. Sodhanaka, my good man, please call Vasantasena's mother here, without frightening her.

SODHANAKA: Yes. (*goes out and comes in with the courtesan's mother*) Come, come this way, madam!

OLD LADY: My daughter has gone to her friend's house to enjoy her youth. But this long-lived gentleman says that I am to go with him as the Judge calls me. Therefore I feel as though a faintness is overpowering me, and my heart trembles with fear. Sir, kindly show me the way to the law-court.

SODHANAKA: Come, come this way, madam.

(*Both walk about*)

SODHANAKA: This is the law-court. May madam enter it.

(*Thereupon both go in*)

OLD LADY (*approaching*): May you be happy, worthy sirs!

JUDGE: Welcome, good lady! Please take a seat.

OLD LADY: Yes. (*sits down*)

SAKARA (*abusively*): So you have come, you old bawd, have you?

JUDGE: Madam, are you the mother of Vasantasena?

OLD LADY: Yes.

JUDGE: And where has Vasantasena gone now?

OLD LADY: To the house of her friend.

JUDGE: What is her friend's name?

OLD LADY (*to herself*): Oh, fie! Oh, fie! This is truly very shameful! (*aloud*) This thing may be asked by a common person, but not by a judge!

JUDGE: Do not feel any shame; it is a legal necessity that is asking you.

SRESHTHIN & KAYASTHA: It is a legal necessity that is putting the question. There is no harm in telling. So please answer.

OLD LADY: What, a legal necessity! If so, then may you gentlemen hear! He is that well-known grandson of the merchant Vinayadatta, and the son of Sagaradatta, who is called the worthy Charudatta, and whose name it is auspicious to invoke—he who dwells in the merchants' quarter. With him my daughter is enjoying the pleasures of youth!

SAKARA: You have heard, gentlemen! Write down these words of hers. My complaint is against Charudatta!

SRESHTHIN & KAYASTHA: There is no harm in the mere fact that Charudatta should be her friend.

JUDGE: This suit now depends on Charudatta giving evidence.

SRESHTHIN & KAYASTHA: So it does.

JUDGE: Dhanadatta, write down, as the first portion of this case, that Vasantasena went to the house of the worthy Charudatta. How now! We have to summon even the worthy Charudatta! Or rather, it is the law that calls him and not us. Sodhanaka, my good man, go and respectfully summon the worthy Charudatta here, gently, without hurry, and without alarming him, telling—'the judge wishes to see you'.

SODHANAKA: As Your Worship orders. (*exit; then entering with* Charudatta) Come, sir, come this way!

CHARUDATTA (*thoughtfully*):—

To tell the truth, this summoning of me, who am very well known to the king as regards my character and lineage, bespeaks a suspicion on his part owing to my present condition of poverty.—

(*reflecting; to himself*)

Could it be that the king found out that Aryaka, whom I met in the road as he was escaping from custody, was sent off by me in my carriage, or did he, having spies for his eyes, come to hear of him as being helped by me, that I am now marching thus as if I were an accused person?

Or rather, why should I think over that? I will forthwith return to the law-court. Sodhanaka, my good man, show the way to the court.

SODHANAKA: Come, sir, come this way.

(*They walk about*)

CHARUDATTA (*with apprehension*): And what is this next?—

The crow is shrieking in a harsh voice; the ministerial servants are calling out for me repeatedly; and my left eye is throbbing violently. These evil omens make me very anxious!

SODHANAKA: Come, sir, come this way, gently and without hurry.

CHARUDATTA (*walking about and looking before him*):—

This crow seated on a withered tree, and facing the sun, is casting his left eye at me. There is no doubt that a dire calamity is to befall upon me.

(*again looking in another direction*) Ah! How is it that here is a snake!—

This huge snake, who was lying asleep across my path with his coiled and swollen belly, is now darting at me in rage, with his eyes fixed on me, his body looking dark like powdered antimony, with his quivering tongue stretched out, and possessing four white fangs.[3]

And moreover, there's this:—

My foot planted on the ground slips, and yet the earth is not very wet. My left eye throbs, and so does my left arm, again and again. And here again is this another crow uttering frequent cries[4]. All this indicates most terrible death: there's no doubt about it.

May the gods grant welfare in all ways!

3. The sight of a serpent confronting a person is considered ominous. It is strange that a snake should be found in a street of such a crowded city as Ujjayini, and that too during day time. But the poet brings in one in his anxiety to bring together all ominous sights.

4. A crow, sitting on a withered tree, facing the sun and uttering harsh cries, anticipates dire calamity.

SODHANAKA: Come, sir, come this way. May Your Honour enter this law-court.

CHARUDATTA (*entering; looking on all sides*): Oh, the supreme grandeur of the court of justice! For here—

Owing to its cruel elements, this royal court of justice looks like an ocean; it has water in the shape of the ministers deeply engaged in thought; it is full of waves and conches in the form of the envoys; it has within its environs sharks and alligators in the form of the spies; it is the abode of ferocious animals in the form of elephants and horses; it is thronged with *Kanka* birds in the form of the litigants shouting various complaints; it is the resort of serpents in the form of the Kayasthas; and it has its shores cut in by the rivers in the form of politics.[5]

Well. (*trying to enter; gesticulates that his head strikes against the door; reflecting*) Alack-a-day! Here's another ill omen!—

My left eye throbs; and the crow is crying; and just now my way was blocked by that snake. May good betide me through fate!

Let me then get in. (*enters*)

JUDGE: Here's that Charudatta, he who—

Possesses a face having a prominent nose and eyes with large corners. Such a face, surely, could never be the abode of having crimes wantonly attributed to it. For in the case of elephants, bullocks, horses and men, the form is never without a corresponding excellence of conduct.

CHARUDATTA: My greetings to the Judges! Well, assessors, how do you do?

JUDGE (*hurriedly*): Welcome to you, noble sir! Sodhanaka, my good man, bring a seat for the noble one.

SODHANAKA (*bringing a seat*): Here's a seat. Please, sir, be seated on it.

5. Sudraka employs animalistic imagery to describe the corrupt court, with elephants and horses serving as the instruments of destruction. The spies are compared with sharks because they furnish secret information to the court; law agents incite people to enter into litigation; Kayasthas(court scribe) are compared with serpents and the general public is like *Kanka* (Herons) shouting their complaints.

(Charudatta *sits*)

SAKARA (*angrily*): Oh! You have come, have you, you woman-killer! Oh, how very fair, how very righteous, is this legal procedure, that a seat should be offered to this woman-killer (*haughtily*): Nevermind; let him have it.

JUDGE: Sir Charudatta, have you any close acquaintance with, or affection, or love, for the daughter of this worthy lady?

CHARUDATTA: Of what lady?

JUDGE: Of this one, here. (*points out the mother of* Vasantasena)

CHARUDATTA (*rising*): Madam, I bow to you!

OLD LADY: May you live long, my son! (*to herself*) Here's that Charudatta! Verily my daughter has most fittingly bestowed her youth!

JUDGE: Sir, the courtesan is your friend?

(Charudatta *feels ashamed*)

SAKARA:—

Having himself killed her for her money, he is now trying to conceal his wicked deed either in shame or out of cowardice. But surely, the king won't conceal it.

SRESHTHIN & KAYASTHA: Noble Charudatta, you may speak out; don't be ashamed. This is simply the legal procedure.

CHARUDATTA (*in shame*): Ye officials, how should I utter such a thing, that a courtesan is my friend? Or rather, herein it is youth that is at fault, and not my character.

JUDGE:—

So leave off all the shame you feel in your mind. Tell the truth: enough of delaying. An equivocal statement would not be accepted here.

Don't be ashamed. It is the law-suit that is putting you the question.

CHARUDATTA: Your Honour, with whom does my law suit lie?

SAKARA (*arrogantly*): Ah you, your suit is with *me*!

CHARUDATTA: A law-suit with you would be extremely unbearable to me.

SAKARA: Ah, you woman-killer, having killed that Vasantasena so beautiful and adorned with hundreds of jewels, you now turn a deceitful rogue and hide it?

CHARUDATTA: You are surely talking nonsense!

JUDGE: Noble Charudatta, never mind him. Tell us the truth. Is that courtesan your friend?

CHARUDATTA: That's so.

JUDGE: Well, where is Vasantasena?

CHARUDATTA: She has gone home.

SRESHTHIN & KAYASTHA: How did she go? When did she go? And who accompanied her when she went?

CHARUDATTA (*to himself*): What! Shall I say that she went unobserved?

SRESHTHIN & KAYASTHA: Sir, tell it please.

CHARUDATTA: She went home. What more should I tell?

SAKARA: You enticed her to my old *Pushpakarandaka* garden and killed her for her money by forcibly strangling her with the noose of your arms. Ah, and now you say that she went home?

CHARUDATTA: Ah, you, talker of nonsense!—

You are not thoroughly wetted by the waters of the clouds in the sky, like the tips of the wings of a *chasha*[6] bird, (and still your face is darkened, without any cause). What you assert is false. For, this face of yours loses its lustre like the lotus in winter.

JUDGE (*aside; to the assessors*):—

Imputing a crime to Charudatta is like weighing the Himalaya mountain, or like crossing the ocean or like capturing the wind.

(*aloud*) He is, indeed the 'noble' Charudatta! How would he do this evil deed? (*repeats the verse beginning 'A face having a prominent nose' et al*)

SAKARA: What! Are you investigating this case with such partiality?

JUDGE: Begone, you fool!—

You, a low caste fellow, are explaining the sense of the Vedas, and yet your tongue does not fall off! You are gazing at the sun at midday,

6. Blue jay, whose wings turn dark when sprinkled over with water.

and still your eyes are not suddenly struck blind! You are putting your hand inside the blazing fire, and yet that hand of yours is not burnt! You are depriving Charudatta of his good name, and still the Earth does not yawn and gulp down your body!

How can the noble Charudatta do an evil deed?—

He who left only the mass of waters in the ocean which was the repository of jewels, and gave away undreamt quantities of money,—how can that high-souled person, the unique abode of all noble qualities, commit, for the sake of money, a crime that is committed only by mean persons?

OLD LADY: You wretch! He who, on that occasion, gave a jewel-necklace, the essence of the four oceans, in return for the golden ornaments deposited with him because they were stolen at night by thieves,—would he now commit this evil deed for the sake of filthy lucre? Oh, my daughter! Come to me, my daughter! (*weeps*)

JUDGE: Worthy Charudatta, did she go walking or in a carriage?

CHARUDATTA: Well, she did not leave in my presence. So I do not know whether she went walking or in a carriage.

VIRAKA (*entering enraged*):—

Nursing a bitter hatred towards Chandanaka for humiliating me with the insult of striking me with foot, as I lay awake brooding over it all this night, it has somehow dawned.

So, then, let me just return to the law-court. (*entering*) Good day, gentlemen!

JUDGE: Ah, here's Viraka, the officer appointed to keep guard over the city. Well, Viraka, what brings you here?

VIRAKA: Oh! Oh! While I was hunting for Aryaka in the tumult following his escape from prison, I grew suspicious about a closed carriage that was passing along, and was going to search it; but then, just as I said: 'Oh, although you have inspected the carriage, I too wish to inspect it,' I was kicked by Chandanaka. Having heard this, may Your Honour decide what to do.

JUDGE: Well, my good man, do you know to whom that carriage belonged?

VIRAKA: To this gentleman, the worthy Charudatta. And the

carriage-driver said that Vasantasena was inside, and that she was being taken to the old garden *Pushpakarandaka* for sporting with him.

SAKARA: You have now once more heard the same story, gentlemen!

JUDGE:—

Alas, here is the moon, shedding a pure light, being eclipsed by Rahu![7] The clear water is being rendered muddy by the falling in of the banks!

Viraka, we'll look into this complaint of yours afterwards. For the present, just mount this horse which is standing outside near the entrance to the court, and go to the *Pushpakarandaka* garden and see if there is any woman lying dead there, or not.

VIRAKA: As Your Honour orders. (*exit; then entering again*) I have just been there. Well, I *did* see a woman's corpse that the jackals were eating up.

SRESHTHIN & KAYASTHA: How did you know that it was a woman's corpse?

VIRAKA: I found it out by the remains of the hair, arms, hands and legs.

JUDGE: Oh, fie upon the intricacy of worldly transactions!—

The more carefully is the case looked into, the darker it appears. Alas! The points of law are very clear; but my judgment finds it difficult to proceed, like a cow in a quagmire.

CHARUDATTA (*to himself*):—

Just as, at its first blooming the bees come in swarm to a flower to suck its honey, so, in times of calamity, a man's troubles multiply when there are weak points.

JUDGE: Noble Charudatta, tell the truth.

CHARUDATTA:—

Whatever a person, who is evil-souled, jealous of another's virtues,

7. One of the navagrahas (nine planets) in Vedic astrology is paired with Ketu. The time of day considered to be under the influence of Rahu is called Rahu kala and is considered inauspicious. In Vedic astronomy, Rahu is considered to be a rogue planet. Also see footnote no. 7, Act IV.

blinded by passion, and desirous of ruining another, tells falsely here, through the evil bent of his very nature,—is all that to be accepted? Is it not to be investigated?

Moreover—

I, who do not even bend down a flowering creeper for plucking its blossoms, and never gather its flowers thus,—how could that I ever kill a weeping woman, seizing her by her long tresses having the dark shine of bees' wings?

SAKARA: You judges, do you conduct your proceedings with partiality, that you allow this wretch, Charudatta, to occupy a seat even now?

JUDGE: Sodhanaka, my good man, do as he says.

(Sodhanaka *does accordingly*)

CHARUDATTA: Consider, O judges, consider what you are doing! (*he then leaves his seat and sits on the bare ground*).

SAKARA (*to himself; dancing in joy*): Heigh-ho! By this the sinful deed done by me has fallen on another's head! So I'll sit where Charudatta, sat (*does so*) Charudatta, look, look at me! Say, then, say that you killed her!

CHARUDATTA: Sirs, judges! (*repeats what he had said before, 'Whatever a person who is evil-souled' et al; then heaving a sigh; to himself*)—

O Maitreya, what is this going on? It will end today in my destruction! O my wife, born in a spotless Brahmana family! O Rohasena! You do not certainly see my calamity! You are ever cheerful, calamities being far off from us; but it is in vain that you do so.

I have sent Maitreya to get news of Rohasena, and also to Vasantasena, to return to her the ornaments given by her to him for making a golden toy-cart out of them. But why is he late?

(*Enter* Vidushaka, *carrying the ornaments*).

VIDUSHAKA: The worthy Charudatta has sent me to Vasantasena, to go to her place, with the ornaments, saying: 'Worthy Maitreya, Vasantasena sent dear Rohasena to his mother, after adorning him with her own ornaments. It was all right for her to give those ornaments, but not for us to accept them. So please return them to her.' So, then, I'll just go to Vasantasena. (*walking about and observing; looking towards the sky*) How! Here is friend Rebhila! Ho, sir Rebhila, why do you look as if you were exceedingly distressed? (*listening*) What do you say? That my dear friend Charudatta has been summoned to appear in the law-court? Ah, then, surely, it could not have been on any ordinary business! (*pondering*) *Well,* I'll go to Vasantasena afterwards. Let me first proceed to the law-court. (*walking about and looking on*) Here is the court of justice; let me then enter it. (*entering*) Hail to you, judges Where is my dear friend?

JUDGE: Here he is!

VIDUSHAKA: Friend, happiness to you!

CHARUDATTA: That will be.

VIDUSHAKA: Why, I hope you are all right?

CHARUDATTA: That, too, will be.

VIDUSHAKA: O friend, why are you looking as if you were extremely distressed? And why have you been summoned here?

CHARUDATTA: Friend,—

Verily, by me, who am cruel and ignorant of the existence of the next world, a woman,—or rather, Rati[8] herself, without any distinction ... The rest this person will tell.

VIDUSHAKA: What, what is it?

CHARUDATTA: (*whispering in his ear*): Thus; like this.

VIDUSHAKA: Who says so?

CHARUDATTA (*points out* Sakara *by a sign*): Why, this poor fellow, who has been the instrument of the god of Death says so of me!

VIDUSHAKA (*aside; to* Charudatta): Why don't you tell them that she went home?

CHARUDATTA: Although told it is not believed in, owing to my poverty.

VIDUSHAKA: O you gentlemen! He who adorned this city of Ujjayini by founding suburbs, and by means of monasteries, gardens, temples, tanks, wells, and sacrificial posts—would he, because he is

8. Hindu goddess of love, carnal desire, lust, passion and sexual pleasure.

poor, commit such an evil deed for the sake of filthy lucre? (*wrathfully*) Oh, you, a whore's son, royal brother-in-law, Samsthanaka, a lost libertine, you scoundrel imputing crimes to people, you monkey decked with numerous golden ornaments, say it in *my* presence! This dear friend of mine, who does not even bend down the *madhavi*[9] creeper in bloom and thereby gather flowers, for fear that perchance its leaves might get torn off the stalk when it is bent,—how could *he* commit such an evil deed, despicable in both the worlds? Wait, you son of a bawd, just wait! I will just break your head into a hundred pieces with this wooden staff, which is as crooked as your heart!

SAKARA (*angrily*): Listen, listen, ye gentlemen! It is with Charudatta that I have a quarrel, or a suit. Why then should this crow-foot-pated fellow break my head into a hundred pieces? You can't do it, you slave-born, knavish Brahmana!

(Vidushaka *raises his wooden staff and repeats what he had said before.* Sakara *gets up in anger, and beats him.* Vidushaka *beats* Sakara *in return. They both beat each other. The ornaments drop down from under* Vidushaka's *armpit*)

SAKARA (*picking them up and looking at them; showing fear*): Look, gentlemen look! These are verily the ornaments of that poor girl! (*pointing to* Charudatta) For the sake of this filthy lucre he strangled and killed her!

(*The* officers *all bend down their faces*)

CHARUDATTA (*aside; to* Vidushaka):—

This group of ornaments, seen at such a time, and dropped from you owing to my ill-luck, will ruin me.

VIDUSHAKA: Well, why don't you tell them the truth?

CHARUDATTA: Friend,—

The royal eye is weak, and it cannot perceive the truth in this matter. If I were to tell, it would be humiliation, and there would be inglorious death.

JUDGE: Alas, Oh, alas!—

9. A virtuous person like Charudatta is harmless and sensitive to the surrounding. Vidushaka argues with the Sakara asking him how he could commit an evil deed.

By the side of Jupiter, powerless and opposed by Mars, there has appeared this another planet, like a comet.

SRESHTHIN & KAYASTHA (*observing; and addressing* Vasantasena's *mother*): Madam, please just inspect these golden ornaments carefully, and find out if they are the same or not.

OLD LADY (*seeing*): They are similar, but they are not the same.

SAKARA: Ha! You old bawd, your eyes have spoken, if your tongue has concealed it!

OLD LADY: Off with you, you wretch!

SRESHTHIN & KAYASTHA: Please answer, without making any mistake, whether these are the same or not.

OLD LADY: Sir, owing to the workman's skill, they captivate my sight, but they are not the same.

JUDGE: Madam, do you recognize these ornaments?

OLD LADY: I say that I do not, I do not, indeed. They are not recognized by me. Or perhaps they might have been made by the same workman.

JUDGE: See, Sreshthin:—

Surely different ornaments resemble each other in form and beauty which are imparted by human art. For all workmen copy the model which they have seen; and perfect resemblance itself is seen on account of their skilled handiwork.

SRESHTHIN & KAYASTHA: Then they belong to the noble Charudatta.

CHARUDATTA: No, surely not!

SRESHTHIN & KAYASTHA: Then to whom do they belong?

CHARUDATTA: To the daughter of this worthy dame here.

SRESHTHIN & KAYASTHA: How did they come to be separated from her person?

CHARUDATTA: They did in this way. Oh, this—

SRESHTHIN & KAYASTHA: Sir Charudatta, you should tell the truth in this matter. For see, see—

Verily, happiness is got by telling the truth. By telling the truth one does not incur sin. *Sa* and *Tya*, only two *akshara* are these, but they are imperishable.[10] Do not hide the truth by telling a lie.

CHARUDATTA: I don't know this much fuss about ornaments; I only know that they were brought here from my house.

SAKARA: First you kill her by enticing her into the garden; and now you hide it by your deceitful cunning!

JUDGE: Noble Charudatta, please tell the truth.—

For now, most assuredly, the heavy lashes of punishment will fall on this delicate body of yours, and so, too, will our wish to acquit you as innocent fall to the ground.

CHARUDATTA:—

I who am born in a family of sinless persons have committed no sin. But if sin is thought to be committed by me, then what avails my being sinless?

(*to himself*) And life is without purpose to me, now that I am separated for ever from Vasantasena. (*aloud*) Gentlemen, what more need I say?—

As is said, by me who am cruel and know neither of the two worlds, a woman, and especially a jewel ... The rest this person will say.

SAKARA: 'Was killed.' Oh! say. you too: 'By me she was killed'

CHARUDATTA: You yourself have said it!

SAKARA: Listen, gentleman, listen! She was killed by him! He himself has removed your doubt! So let death be inflicted upon this penniless Charudatta.

JUDGE: Sodhanaka, it shall be as the king's brother-in-law says. Well, guards, arrest this Charudatta!

(*The guards arrest him*)

OLD LADY: Be kind, gentlemen, please be merciful. (*repeats what she had said before, 'He who on that occasion...because it was stolen by thieves ...' et al*) So, if my daughter has been killed, well, she has been killed. But let my long-lived one, Charudatta, live. Moreover, a suit lies between the complainant and the defendant; and *I* am the real

10. Satya literally meaning 'truth' in Sanskrit is made up of two syllables only 'Sa' and 'Tya'. Sreshthin & Kayastha remind Charudatta to tell the truth which is potent.

complainant in this matter. So please release him.

SAKARA: Get away, you born slave, off with you! What have you to do with him?

JUDGE: Madam, you may go now. Guards, please show her out.

OLD LADY: O my dear, my child! (*exit weeping*)

SAKARA (*to himself*): I have done to him something which is worthy of myself! Now I'll go. (*exit*)

JUDGE: Sir Charudatta, we are authorized merely to give our decision in a case; but the rest depends on the king. Still, O Sodhanaka, let King Palaka be informed in these words:—

'He is a sinner. But Manu[11] has laid down that a Brahmana is not to be killed; so he might be banished from this country together with all his property intact.'

SODHANAKA: As Your Honour orders. (*goes out and again comes back; tearfully*) Gentlemen, I had gone there. But King Palaka says—'Tie those very ornaments round the neck of him who killed Vasantasena for the sake of a paltry sum of money; and, beating the drum, take him to the southern burning-ground and have him impaled there. Whoever else commits a similar crime will be visited with the same punishment, accompanied by great disgrace.'

CHARUDATTA: Ah! How thoughtlessly has King Palaka acted! Or rather—

It is but right that kings, who are the protectors of the earth, are reduced to pitiable plight, when pushed into such a fire of legal trials by their counsellors.—

(*pointing towards the* Judge) By such men, who would even believe that a crow is white, and who disgrace the king's judicial administration, thousands of sinless persons are being, and have been, killed.

Friend Maitreya, please go, and at my request give, for the last time, my salute to my dear mother. And please look after my son, Rohasena.

VIDUSHAKA: When the root itself is cut off, how can the tree be looked after?

11. Laws of Manu as mentioned in *Manusmriti*.

CHARUDATTA: Nay, say not so.—

The son is the physical substitute of persons who have departed to the other world. So please bestow on Rohasena that very affection which you had for me.

VIDUSHAKA: O friend, shall I be able to support life separated from you, having been your dear friend so long?

CHARUDATTA: At any cost, please let me have a sight of Rohasena.

VIDUSHAKA: Quite so; that is all right.

JUDGE: Well, Sodhanaka, let this fellow be removed from here.

(Sodhanaka *does so*)

JUDGE: Ho! Who is there? Let orders be issued to the Chandalas.

(*Exit all the* officers *leaving alone* Charudatta)

SODHANAKA: Sir, please come this way.

CHARUDATTA (*sorrowfully recites the verse beginning 'O Maitreya, what is this today' et al; looking towards the sky*):—

After my case had been investigated, by means of the ordeals of poison, water, balance, and fire,[12] then with justice you might have today ordered a saw to be used on this body of mine. But as you are killing me, a Brahmana, relying simply on the words of my enemy, you will surely fall into hell, along with your sons and grandsons!

Here I am coming!

(*Exit all*)

Thus ends Act IX, entitled

'The Trial'

12. As per Hindu mythology, there are various modes to test one's innocence by an ordeal, the four mentioned here are—*poison*: making a man taste poison without his being affected or harmed by it; *water*: undergoing the ordeal by water involves submerging under water and staying there until another person has recovered an arrow discharged at the time of submerging. Innocence is proven if the runner is back in time; *balance*: by placing him in a scale-pan with equal weight put on the other, and then seeing that his pan rises up if innocent; and *fire*: by observing that he remains unburnt in fire for a certain space of time, after the performance of some preliminary rites.

ACT X

(*Enter* Charudatta, *accompanied by two* Chandalas)

CHANDALAS:—

What then! Seek not the reason. It is this, that we are experts in chaining and leading a recently condemned criminal to the gallows, and are skilled in cutting off his head or impaling him without loss of time.

Make way, gentlemen, make way! Here's the worthy Charudatta—

On whom garlands are put of *karavira*[1] flowers and who is in the custody of us two, his executioners; he is now slowly losing lustre, like a lamp with its oil exhausted.

CHARUDATTA (*mournfully*):—

These crows, cawing harshly nearby, are expecting to feed on my body as though it were an oblation; my body which is drenched with tears, has its limbs soiled with dust, is covered with garlands of funeral flowers, and is besmeared with red sandal-paste.[2]

CHANDALAS: Make way, gentlemen, make way!—

Why should you look at this good man being hewn[3] asunder by the edge of Death's axe, this tree in the shape of a righteous man, being the residence of birds in the form of virtuous persons?

Come, O Charudatta, come away!

CHARUDATTA: The workings of a person's destiny are verily inscrutable, since I have been reduced to such a condition—

By the prints of the extended hand dipped in red-sandal paste, impressed over all my limbs and covered over with rice flour and sesamum-powder, I, a man, am turned into an animal about to be

1. Sweet-scented oleander.
2. The paraphernalia and circumstances of a public execution according to Hindu fashions are interestingly described here; the low official attendance shows that the people were as easily managed then as at any subsequent period. The character of the executioner corresponds precisely with that of the Roman carnifex, and in like manner the place of execution is the public cemetery, or place of burning the dead. The criminal is dressed as a victim with very classical decorations.
3. Strike forcibly with a sharp cutting object.

slaughtered.

(*looking before him*) Oh, the commonsense of the people! (*mournfully*)—

Here these citizens, seeing the state to which I have now fallen, are saying 'Fie upon this moral lot', with tears welling up in their eyes, and being unable to save me, are blessing me with words 'May you go to heaven!'

CHANDALAS: Make way, sirs, make way! What are you seeing?—

The *Indradhwaj*[4] being carried away, the delivery of a cow, the falling of a star, and the execution of a virtuous man,—these four ought not to be looked at!

FIRST CHANDALA: Eh, Ahinta[5] see, see!—

When the chief man of the city is about to be slain by the fiat of the Fate, is the sky weeping, or is the thunderbolt falling, though there are no clouds?

SECOND CHANDALA: O Goha[5]—

The sky is not weeping, nor is the thunderbolt falling in the absence of clouds. But these are the tears, flowing in torrents, from the clouds in the form of the assembled ladies.

Moreover—

As the condemned man is being led on, the dust does not rise up from the streets, being sprinkled over with the tears of all people who are weeping.

CHARUDATTA (*observing; mournfully*):—

And these ladies in their mansions, putting forth their faces through the lower halves of the windows, are saying, with reference to me, 'Alas, O Charudatta!' and are shedding tears profusely.

CHANDALAS: Come, O Charudatta, come along! Here's a proclamation-station. Beat the drum, and repeat aloud the proclamation!

CHANDALAS: Listen gentlemen, listen! This is the grandson of the merchant Vinayadatta, and the son of Sagaradatta, who is called the worthy Charudatta. By him, who has done an evil deed, the courtesan Vasantasena was enticed into the old deserted *Pushpakarandaka* garden

4. Indra's pole. See Glossary.
5. Names of the two chandalas.

for the sake of a trifle, her money, and was there killed by being forcibly strangled with the noose of his arms; then he was caught with the booty, and himself also confessed his crime. Thereupon King Palaka ordered us to slay him. If anyone else commits a similar crime, which is good neither for this world nor for the next,[6] him too King Palaka will condemn similarly to death.

CHARUDATTA (*with dejection; to himself*):—

My family, which was formerly sanctified by the performance of hundreds of sacrifices, and made illustrious by the chanting of Vedic hymns in crowded halls in sacrificial assemblies, is now being loudly named in the proclamation and polluted by such persons, unholy and quite different from my ancestors, as I am now a man condemned to death.

(*shuddering and covering up his ears*) Oh, dear Vasantasena!—

O you, with teeth white like the pure rays of the moon and with the lower lip resembling the beautiful coral, having drunk the nectar from your mouth, how can I, who am so helpless, now swallow the poison in the form of ignominy?

CHANDALAS: Make way, sirs, make way!—

This store of the jewels of virtues, this bridge for the good to cross the ocean of their miseries, this ornament, though not one of gold, this person is today being taken away from the city.

Moreover—

Verily, in this world, all interest themselves in other persons when these are in easy circumstances. But it is difficult to find one who will show kindness to men who have fallen.

CHARUDATTA (*looking on all sides*):—

These friends of mine, covering up their faces with the ends of their garments, are going further away from me. Even a stranger behaves like a close relative to a person in normal circumstances but no one befriends a man whose circumstances are reversed.

CHANDALAS: We have turned away the crowd. The high road is now clear. So bring him up, with the signs of the condemned put on.

6. The idea of Loka (this world) and Parloka (celestial world) in Hindu worldview.

(Charudatta *sighs and recites the verse beginning 'O Maitreya, what is this today' et al*)

(*Offstage*)

Alas, father! Alas, dear friend!

CHARUDATTA (*listening; mournfully*): O you, the chief of your caste, I wish to ask a favour at your hands.

CHANDALAS: What! You would take a favour from *our* hands?

CHARUDATTA: Heavens forbid! Still, a Chandala is not like the ill-behaved King Palaka, who acts without due investigation. So, I request you for getting a last sight of my son's face, so that I might easily obtain heaven.

CHANDALAS: You may do so.

(*Offstage*)

Alas, Alas, father!

(Charudatta, *hearing it, mournfully repeats 'O you the chief of your caste' et al)*

CHANDALAS: Ye citizens, make room for a moment. Let this worthy Charudatta have a look at his son's face. (*in the direction of* Charudatta's *son*) Sir, this way, in this direction! Come, boy, come up here!

(*Enter* Vidushaka *with* Rohasena)

VIDUSHAKA: Make haste, my sweet-faced one, make haste; your father is being taken to be killed!

ROHASENA: Oh, father! Oh, father!

VIDUSHAKA: Oh, dear friend! Where shall I see you again?

CHARUDATTA (*seeing his son and his friend*): Ah, my son! Ah, Maitreya! (*mournfully*) Oh! Alas!—

Verily I shall have to remain thirsty for a long time in the next world; for very small, indeed, would be this libation of water offered to me after death as the boy's hands are so small.[7]

What shall I bequeath to my son? (*looking over his own body and*

7. Rohasena being a boy, the water he will offer by his tiny hands would not be sufficient to allay Charudatta's thirst. It is a Hindu belief that the departed forefathers subsist on the food and water offered to them by their descendants.

seeing his sacred thread) Ah, yes! Here's a thing which I do possess—

Is an ornament to the Brahmanas, though not made of pearls or of gold, and through which oblations are offered to the gods and the manes[8].

(*He hands over his sacred thread*)

FIRST CHANDALA: Come, O Charudatta, come along!

SECOND CHANDALA: Ah, you are calling the worthy Charudatta without his title of Arya[9]? See, you!—

Unchecked in its course, like a female colt broken loose from the reins, Fate verily comes thundering along in pursuit of a man, by day and by night, in prosperity as well as in adversity.

Moreover—

Have his titles become empty? Should we not bow down our heads to him? We should. Does not the moon, though eclipsed by Rahu, deserve to be saluted by the people in general?

ROHASENA: Oh, you Chandalas, where are you taking my father?

CHARUDATTA: Child!—

Bearing the garland of *karavira* flowers round my neck, the iron spike on my shoulder and grief in my heart, I am today going to the place of execution, like the goat at a sacrifice being led to the place of immolation.

CHANDALAS: Boy!—

Verily we are not to be called Chandalas, though we have been born in a Chandala family. Those who prosecute a virtuous man,—*they* are the sinners, and *they* the Chandalas.

ROHASENA: Then why are you killing my father?

CHANDALAS: My long-lived boy, it is the king's command that commits the wrong is to blame, and not we two.

ROHASENA: Kill me, and let my father go!

CHANDALAS: My long-lived child, may you that speak thus, live long!

CHARUDATTA (*embracing his weeping son around the neck*):—

8. Worshipped as beneficent *Pitris* (ancestoral spirits).
9. An honorific used in ancient India. Here, worthy.

This is the all-in-all of love, equally accessible to both, the rich and the poor! It is a soothing ointment to the heart, though containing neither sandal nor *usira*[10] paste.

(*He again recites the verse beginning 'Bearing round* my *neck' et al; then he observes, and again repeats to himself the verse beginning 'And these with their faces covered up with the ends of their garments et al*)

VIDUSHAKA: You gentle ones, let go my dear friend Charudatta, and kill me instead!

CHARUDATTA: Heavens forbid! (*observing; to himself*) Now I have learnt it! (*he recites again 'Of one in happy circumstances' et al; then aloud; he recites again the verse beginning 'And these ladies in their mansions' et al*)

CHANDALAS: Make way, sirs, make way—

Why should you throng to look at a virtuous man who has lost all hope of life owing to ignominy, like a golden jar sinking in a well when its rope is snapped?

(Charudatta *mournfully recites the verse beginning 'O you with teeth. pure like the rays of the moon' et al*)

CHANDALAS: Eh, proclaim it once more!

(*The* Chandalas *do so*)

CHARUDATTA:—

I have already been reduced to a despicable state, brought on by adversity, having for its climax the loss of my life in this fashion. And this proclamation cuts me to the quick that I should have now to hear that that beloved of mine was killed by me!

(*Enter* Sthavaraka *in fetters and imprisoned in the palace*)

STHAVARAKA (*listening to the proclamation; distressfully*): How now! They are killing the innocent Charudatta! And my master has secured me with chains! Never mind, I will shout out.—Listen, sirs, listen! It so happened that I, the sinful wretch, took Vasantasena to the old *Pushpakarandaka* garden, owing to an interchange of carriages. Then my master killed her by violently strangling her with the noose of his arms, because she did not love him; this noble person didn't do

10. Fragrant root of a herb.

it. How! Owing to the great distance nobody hears me! Then what shall I do? Shall I throw myself down? (*reflecting*) If I do so, the worthy Charudatta would not be killed. All right. I will throw myself down through this old window, from this elephant-corniced terrace of the palace. Better that I should be dead than the noble Charudatta, who is the tree affording shelter to the birds in the form of all well-born youths. If I die thus, I shall have obtained heaven! (*thereupon he throws himself down*) Wonderful! I have not been killed, and these my fettering chains have snapped asunder! Now I shall find out where that proclamation of the Chandalas comes from. (*observes and approaches*) Oh, you Chandalas, make room, make room!

CHANDALAS: Eh, who is that wants room?

(Cheta *repeats what he had said before, 'Listen' et al*)

CHARUDATTA: Ah!—

Who is this that has sprung up at such a time, when I am held fast in the noose of Death, like the *Drona* cloud[11] when the corn is suffering from drought?

Well, Sirs, have you heard?—

I am not afraid of death; only my fair name is tarnished and this is what pains me. Death to me if proved, will be as pure as the birth of a son.

Moreover—

By that mean fellow, Sakara, whose enemy I never was and who is extremely narrow-minded, I was charged with crime—though he himself was the criminal—as with a poison-besmeared arrow.

CHANDALAS: Sthavaraka, are you telling the truth?

STHAVARAKA: Yes, the truth. And he also put me in chains linked together, and kept me, imprisoned in the elephant-corniced terrace of his palace, saying 'you will not tell anyone.'

SAKARA (*entering; gleefully*):—

In my palace I have dined on rice, and meat seasoned with pungent and sour sauces, and vegetables, soup and fish, and rice-cakes and treacled rice.

11. A particular class of clouds discharging rain plentifully and producing luxuriant crops.

(*listening*) It seems to be the voice of a Chandala speaking, as it has the ring of cracked bell-metal. And, since this loud noise of drums and kettledrums beaten at the time of execution is heard, I guess that the penniless Charudatta is being taken to the place of execution. So I shall go and have a good look at it. The destruction of an enemy, in sooth, affords my heart very great pleasure. And I have heard that whoever witnesses the killing of an enemy will not suffer from any eye-disease in his next birth. I have indeed managed to bring about the ruin of that pauper Charudatta, after having sought a broad way out, like a warm stationed in the inside of the root of a poisonous herb. Now I'll ascend to the elephant-corniced upper terrace of my palace and from there witness my own exploit.

(*does so; seeing*) Ha, ha! There is so large a crowd of people when this beggar Charudatta is being taken to the execution-ground; how large then would it be when a mighty high personage like myself would be taken to be executed? (*peering*) How! Here he is, decked like a newly trained bullock, being taken in the southern direction! But why is it that the proclamation was made in the neighbourhood of the elephant-corniced terrace of my palace, and stopped suddenly? (*observing*) How! The slave Sthavaraka, too, is not to be seen here! Could he have gone from this place and divulged the facts? Let me then first seek him out.

(*Descends and approaches*)

STHAVARAKA (*seeing him*): Gentlemen, here he comes!

CHANDALAS:—

Move off and make way! Shut the doors and remain silent! For here comes this mad bull having sharp horns in the form of his arrogant deeds!

SAKARA: Ho, there! Make way, make way! (*approaching*) My son, Sthavaraka, my slave, come, let us be off!

STHAVARAKA: Oh, you villain! You were not satisfied by the murder of Vasantasena, and now you want to murder the noble Charudatta also, the wish-granting tree of all supplicants!

SAKARA: Never shall I, who am like a pot filled with jewels, murder a woman for the sake of her money!

ALL: Oh, but she was murdered by *you*, and not by the worthy

Charudatta!

SAKARA: Who says so?

ALL (*pointing to* Sthavaraka): Why, this good man!

SAKARA (*alarmed; to himself*): Hang it! How is it that I failed to properly secure the slave Sthavaraka? For, he is the sole witness of my crime. (*pondering*) Well, I'll manage it thus. (*aloud*) It is a lie, gentlemen! For see, this slave was caught by me as he was stealing my gold, and beaten and cudgelled and imprisoned. So, having become my enemy, whatever he tells you—could it be the truth? (*aside; gives a golden bangle to* Cheta; *whispering*) Sthavaraka, my son, my slave, take this and contradict what you have said.

STHAVARAKA (*taking it*): Look, masters, look! Oh, here he is tempting me with gold!

SAKARA (*snatching away the bangle*): Here's that golden ornament for which he was put in chains! (*angrily*) Oh you Chandalas, I really did beat and cudgel him as he was stealing my gold when appointed to guard my treasury. And, if you don't believe me, you may just inspect his back.

CHANDALAS (*seeing*): What he says is right. What will a slave when exasperated not prattle?

CHETA: Alas! Slavery is such a cursed thing that nobody believes the truth told by a slave. (*mournfully*) Sir Charudatta, only thus far lay my power! (*he prostrates himself at his feet*)

CHARUDATTA (*mournfully*):—

Rise! O you who took pity on a fallen innocent man, who came to me as a disinterested friend, and who are righteous by nature. Although you made a very great effort to set me free, Fate does not agree to it. What is it that you have not done for me today?

CHANDALAS: Sir, thrash this slave and turn him out.

SAKARA: Here, off with you! (*he drives him out*) Well, Chandalas, why do you delay? Kill him!

CHANDALAS: If you are in such a hurry, then you may kill him yourself!

ROHASENA: O Chandalas, kill me; release my father!

SAKARA: Kill him together with his son!

CHARUDATTA: Everything is possible in the case of this fool; so, my son, go to your mother.

ROHASENA: What am I to do by going away?

CHARUDATTA:—

Dear child, this very day you should go to reside in a hermitage in a forest, along with your mother, lest you too, my son, fare similarly, through the crime of your father.

So, friend Maitreya, take him and go.

VIDUSHAKA: O friend, have you so known it as to think that I could remain alive without you?

CHARUDATTA: Friend, your life is in your own power, and it is not right for you to give it up.

VIDUSHAKA (*to himself*): That is surely true; but still, I am unable to remain alive after being separated from my dear friend. So, I will give this boy into his wife's care and myself follow my dear friend by giving up my life. (*aloud*) O friend, I shall immediately take him back. (*he embraces him around the neck and falls at his feet*)

(Rohasena, *too, weeps and prostrates himself*)

SAKARA: Eh, do I not say that you should kill Charudatta along with his son?

(Charudatta *exhibits alarm*)

CHANDALAS: The king's order to us is not such to the effect that we should kill Charudatta along with his son. Here, go away, boy, off with you! (*they show him out*) Here's the third proclamation-station. Beat the drum! (*they proclaim once more*)

SAKARA (*to himself*): How! These citizens do not believe it! (*aloud*) Oh, Charudatta, you fellow! This multitude of citizens do not believe it. So speak it with your own tongue: 'I killed Vasantasena.'

(Charudatta *remains silent*)

SAKARA: Ho! Chandala, man, this fellow, Charudatta, doesn't speak. So make him speak by repeatedly striking him with that broken piece of old bamboo, your drum-stick.

CHANDALAS (*raising his hand for a blow*): Speak, O Charudatta!

CHARUDATTA (*mournfully*):—

Having fallen in this fathomless big ocean of calamity, I feel neither terror nor sorrow in my mind. But only one thing—the fire of popular ignominy, burns me, when now I am asked to say—'I killed my beloved.'

(Sakara *again says as before)*

CHARUDATTA: Oh, ye citizens! (*he recites again the verse beginning 'By my cruel self' et al*).

SAKARA: 'Was killed.'

CHARUDATTA: So be it!

FIRST CHANDALA: Oh, it is your turn today to kill.

SECOND CHANDALA: Oh, it is yours!

FIRST CHANDALA: Let us reckon. (*they write or draw lines in many ways*) Ah, if it is my turn, then let it stand over for a short while.

SECOND CHANDALA: What for?

FIRST CHANDALA: I was told by my father when about to depart to heaven, thus: 'Viraka, my son, whenever it will be your turn to kill, you should never kill the condemned man at once.'

SECOND CHANDALA: For what reason?

FIRST CHANDALA: Perchance some kind gentleman might set the condemned man at liberty by paying a ransom. Perhaps a son might be born to the king, and on account of that great joyous festival, all the condemned men may be released. Perchance an elephant might break loose from his chains, and in that commotion the condemned criminal might escape. Or sometimes there might be a change of kings, and then all the condemned men may be set at large.

SAKARA: What, what's that? A change of kings?

CHANDALAS: Eh, let us reckon by writing out our turns of killing.

SAKARA: You there, kill Charudatta quickly! (*having said so he, with* Cheta, *moves away and remains in a corner*)

CHANDALAS: Noble Charudatta, it is verily the king's command that is guilty towards you, and not we, Chandalas. So please think of whatever you have to think of.

CHARUDATTA:—

Although through ill luck I am somehow condemned by the words of persons in high authority, still, if my righteousness possesses power, then let that very same beloved of mine, residing in Indra's heaven,—or wherever she may be,—remove this stigma on me, by virtue of her own good nature.

Well, where am I to go now?

CHANDALAS (*pointing*): Well, here you observe this southern cemetery[12] at whose very sight the condemned men at once take fright and die. See, see—

Jackals with elongated bodies are pulling at the lower half of a corpse[13] which is dangling there, while the other half still remains on the stake, like the grinning mask of loud laughter.

CHARUDATTA: Alas! I am undone, unlucky that I am! (*sits down in agony*)

SAKARA: I won't go just yet. Let me first see Charudatta being killed. (*walking about and seeing*) How! He has sat down!

CHANDALAS: Charudatta, are you afraid?

CHARUDATTA (*getting up at once*): You fool! (*he recites again the verses 'I am not afraid of death; only my good name is soiled' et al*)

CHANDALAS: Noble Charudatta, even the sun and the moon, who are denizens of heaven, are overtaken by calamity, what then of common people, men who are afraid of death? In this world, anyone who rises does fall, and one who falls does rise again.—

The dead body, rising and again falling, exhibits the manner of the casting-off of a garment. So, keeping these things in your mind, you should steel yourself against death.

(*to the second* Chandala) Here's the fourth proclamation-station! Let us, therefore, proclaim it aloud, (*they again proclaim as before*)

CHARUDATTA: Oh, my beloved Vasantasena! (*he again recites the verse beginning 'O you with teeth pure like the* rays *of the moon' et al)*

(Enter, in a hurry Vasantasena *and* the Mendicant)

MENDICANT: Wonderful! I am indeed blessed by my ascetic

12. Place of execution.

13. Jackals with their outstretched bodies trying to reach the dangling bodies.

state, since I am leading Vasantasena after having revived her who had swooned in an undesirable locality—O, worshipper of Buddha, where shall I take you?

VASANTASENA: To the house of the worthy Charudatta, of course! Kindly gladden me with his sight, like a night-lotus with that of the moon!

MENDICANT (*to himself*): By what route shall I proceed? (*pondering*) I will go by the high-road. Come, O worshipper of Buddha! Here is the high-road. (*listening*) What, I wonder, might be the cause of this mighty uproar that I hear on the high-road?

VASANTASENA (*peering before her*): How! There's a big crowd of people in our front! Sir, please just find out what it may be. Ujjayini seems to have raised itself in one spot, as though the earth had become burdened with an uneven load!

CHANDALAS: And here is the last proclamation-station. So, strike the drum and repeat aloud the proclamation! (*doing accordingly*) O Charudatta, get ready! Do not fear; you will quickly end!

CHARUDATTA: Ye gods in heaven!

MENDICANT (*hearing it; excitedly*): O worshipper of Buddha! Charudatta is being taken for being killed on a charge of having murdered you.

VASANTASENA (*in agitation*): Oh, alas! Oh, alas! How is it that the worthy Charudatta is being killed on account of me, the unlucky one! Sir, please show me the way very quickly.

MENDICANT: Make haste, O worshipper of Buddha, make haste and comfort the worthy Charudatta while he still lives! Make way, sirs, make way!

VASANTASENA: Make room, make room!

CHANDALAS: Noble Charudatta, it is our master's command that is guilty here. So remember whatever you have to remember.

CHARUDATTA: What more need I say? (*recites the verse beginning 'Although I am deemed a sinner' et al*)

CHANDALAS (*drawing his sword*): Noble Charudatta, lie flat and keep straight. We will complete you with one stroke and send you to heaven.

(Charudatta *lies as asked*)

CHANDALAS (*about to strike; gesticulating that the sword drops from his hand*): Ha! How's that?—

Although I fiercely drew it by its hilt and grasped it in my fist, why has the sword, terrible like the thunderbolt, fallen on the ground?

Since this has occurred, I infer that the worthy Charudatta won't be killed after all! O divine Durga[14], You that dwellest on the Sahya! Have mercy! If Charudatta happens to be saved, then forsooth the whole race of the Chandalas will have been blessed by you!

SECOND CHANDALA: Let us do as we were ordered.

FIRST CHANDALA: Certainly. Let us do so.

(*Thereupon both prepare to fix* Charudatta *on the stake*)

(Charudatta *again recites the verse beginning 'Although I am deemed a sinner' et al*)

MENDICANT & VASANTASENA (*observing*): Forbear, sirs, forbear! Gentlemen here I am, that same unlucky woman for whose sake he is being killed!

CHANDALAS (*seeing*):—

And who may this lady be, with her ample tresses fallen over her shoulders, that is quickly coming in this direction, with her hand raised, and saying 'Forbear, forbear'?

VASANTASENA: Noble Charudatta, what does this mean? (*she falls on his breast*)

MENDICANT: Noble Charudatta, what may this be? (*he falls at his feet*)

CHANDALAS (*moving off in alarm*): How! She is Vasantasena! How very lucky that we did not kill an innocent person!

MENDICANT (*rising*): Man, does Charudatta live?

CHANDALAS: He lives a hundred years!

VASANTASENA (*joyfully*): I am restored to life again!

CHANDALAS: So, then, let us first report this turn of events to the

14. The goddess Durga in her fierce form, known as Chandika, the patron deity of the Chandalas, Bhils, and other such fierce tribes. Here she is called Sahya, which shows that the Chandalas came from the mountain Sahya.

king who is in the sacrificial enclosure. (*they start to go*)

SAKARA (*seeing* Vasantasena; *in alarm*): Heavens! By whom was this born slave-girl brought back to life? Now my life has ended! Well, let me take to my heels. (*he runs* away)

CHANDALAS (*approaching*): Oh, the king's order to us was to this effect—'Kill him by whom she was murdered.' So let us go in pursuit of the royal brother-in-law himself. (*exit*)

CHARUDATTA (*full of wonder*):—

While the weapon was raised and I was in the jaws of death, who is this lady that has come up, like a shower from the Drona cloud, coming on the crop withering owing to drought?—

(*Then, having observed*)

Is this a second Vasantasena? Or has that same thus come here from heaven? Or is my wandering mind seeing her in illusion? Or can it be that Vasantasena never died at all, and that this is the same?

Or again:—

Has she come back again from heaven with a desire to restore me to life? Or is she some different lady, resembling her in appearance, that has come here?

VASANTASENA (*tearfully rising, and falling at his feet*): Noble Charudatta, it is my own sinful self, for whose sake you were reduced to this unmerited plight.

(*Offstage*)

Wonder, oh, wonder Vasantasena is alive! (*all repeat these words*)

CHARUDATTA (*hears, immediately rises, and exhibits pleasure at* Vasantasena's *touch; Still keeping his eyes closed, and in a voice faltering in ecstasy*): Beloved mine, are you really Vasantasena?

VASANTASENA: The very same, my own unlucky self!

CHARUDATTA (*observing; joyfully*): How! She is Vasantasena and none else! (*delightedly*)—

Bathing your breasts with the streams of the water of tears, whence did you come, like the divine lore, when I was in the grasp of Death?

Beloved Vasantasena!—

This body of mine, which was about to be destroyed on your account, was saved by yourself from that fate. Oh, the power of union with one's beloved person! Else who, being dead, could ever be brought back to life again?

And, O my dear, see, moreover:—

By the arrival of you, my beloved, that same red garment has become like a bridegroom's attire, and this garland decks me as though I were a bridegroom. And similarly, these sounds of the drum of execution now resemble those of nuptial drums.

VASANTASENA: What is this that your noble self, in your extreme generosity, has brought on yourself?

CHARUDATTA: Beloved, under a charge that you were killed by me,—

I was very nearly put to death by that powerful enemy of mine, who had from the very beginning conceived hatred against me and who is to fall in hell.

VASANTASENA (*closing her ears*) Heavens forbid! It was that royal brother-in-law who killed me.

CHARUDATTA (*seeing the* Mendicant): And who is this, too?

VASANTASENA: I was killed by that wicked man; but this gentleman restored me to life.

CHARUDATTA: Who are you that thus befriended her disinterestedly?

MENDICANT: Your worthy self does not recognize me. I am that same person, named Samvahaka, employed as the Shampooer of your noble self. I was seized by gamblers, but this worshipper of Buddha bought me off giving an ornament as my ransom, simply because I belonged to Your Honour. And through the disgust which I then felt for gambling, I turned a Buddhist, mendicant. This worthy lady had gone to the old *Pushpakarandaka* garden owing to an interchange of carriages; and she was forcibly strangled by that wicked fellow with the noose of his arms, because she didn't think highly of him. Later, I found her.—

(*An uproar offstage*)

Victory to the bull-bannered deity Siva,[15] the destroyer of Daksha's sacrifice! After that, victory to the six-faced Karttikeya, the cleaver and the enemy of the Krauncha mountain! And after that, victory to Aryaka, who has killed his mighty foe and conquered the whole vast expanse of the earth, having the white Kailasa mountain for its white banner!

(*Entering suddenly*)

SARVILAKA:—

Ye citizens! Having killed that wicked King Palaka, and having immediately installed that Aryaka as sovereign of his kingdom, I have reverently received with bent head his command as if it was the cast-off garland of worship, and will forthwith release Charudatta, who is in danger.—

Having killed that enemy who lacked both strength and wise counsel, and having afterwards inspired confidence among the citizens through our superior prowess, we have acquired the whole enemy's kingdom, with sovereignty over the world, as if it were Indra's own kingdom!

(*looking before him*) Yes. He must be just there where this crowd of citizens is to be seen. I hope that this enterprise of King Aryaka will be crowned with success by the saving the life of the worthy Charudatta! (*walking very swiftly towards them*)

Away, you rash people! (*observing joyfully*) Surely, Charudatta is alive, and so is Vasantasena! The heart's desire of my master is now fulfilled, indeed!—

Aye! By great good luck I am beholding after a long time, like the moon freed from eclipse and splendidly united with moonlight, Charudatta who has crossed the boundless and vast ocean of misery by the aid of his best beloved, possessing virtues and high moral character, as by means of a boat held fast by ropes and well-steered.

But then, how possibly shall I approach him, as I have committed the great sin of theft, in his house? Or why, straightforwardness is

15. The bull-emblemed god, Siva. The destroyer of the sacrifice of Daksha, his father-in-law. He first glorifies Siva, and afterwards his friend Aryaka, who also has destroyed the sacrifice of Palaka, killing him, as Daksha was killed by Siva.

becoming on all occasions. (*coming forward; folding his hands*) Sir Charudatta!

CHARUDATTA: Ah, who are you?

SARVILAKA:—

I am that same man who broke into your house and stole that deposit; having committed that great sin, I now throw myself on your mercy!

CHARUDATTA: Friend, say not so! It was a *kindness* which you did to me! (*embraces him*)

SARVILAKA: Moreover—

The wicked-souled Palaka, who was staying in the sacrificial enclosure, was slaughtered like a sacrificial animal by Aryaka, of noble deeds, thus maintaining the greatness of his family and his honour.

CHARUDATTA: What!

SARVILAKA:—

He, who formerly drove in your chariot and threw himself on your mercy, has today slain Palaka like a sacrificial animal, in the course of a sacrifice that was being performed.

CHARUDATTA: Sarvilaka, do you mean that same person, named Aryaka, whom you helped to escape and who was brought by Palaka from his hamlet and imprisoned in a secret dungeon without any cause?

SARVILAKA: It is, as Your Honour says.

CHARUDATTA: Then it is gratifying news to me, gratifying indeed!

SARVILAKA: Immediately on his being enthroned in Ujjayini, your friend Aryaka has bestowed on you the kingdom of Kusavati on the banks of the Vena river. So please honour by your acceptance this first request of your friend. (*turning round*) You there! Bring here that sinful scoundrel, the royal brother-in law!

(*Offstage*)

As Sarvilaka orders!

SARVILAKA: Sir, King Aryaka has indeed requested you in these words: 'This kingdom was acquired by me through your goodness. It is therefore at your disposal.'

CHARUDATTA: What! The kingdom was acquired through my goodness?

(*Offstage*)

Come, you brother-in-law of the king, come along! Taste the fruit of your own evil deeds!

(*Enter* Sakara, *led in by the guard with his hands tied behind him*)

SAKARA: Wonderful!—

Although I had escaped to such a long distance, like a donkey broken loose from his reins, I have indeed been brought back a captive, as if I were an untamed wild dog.

(*looking in all directions*) The king's brother-in-law is now surrounded on all sides by these guards. So then, being helpless, whose protection shall I now seek? (*pondering*) Yes. I will approach that very same person, who is kind to those that seek his protection. (*thinking so; he goes forth*) Noble Charudatta, save me! save me (*he throws himself at his feet*)

(*Offstage*)

Let him off, Sir Charudatta, let him off! We will kill him.

SAKARA (*to* Charudatta): Save me, sir, you who are the refuge of the helpless!

CHARUDATTA (*mercifully*): Oh, pity! I promise safety to you who seek my protection! You are safe!

SARVILAKA (*impatiently*): Oh! Drag him off from Charudatta's side! (*to* Charudatta) Well, sir, tell us what should be done with this sinful wretch.—

Let Chandalas bind him tight and drag him away. And then, should he be devoured by dogs, or lie on the stake, or be cut up with a saw?

CHARUDATTA: Well, will you do what I tell?

SARVILAKA: What doubt is there about that?

SAKARA: Master, Charudatta, I beseech your protection! So save me, Oh, save me! Do what behoves you! I will never again behave similarly!

(*Offstage; the citizens speaking*)

Kill him! Why is the sinner allowed to live?

(Vasantasena *removes the garland, the badge of execution, from* Charudatta's *neck and throws it on* Sakara)

SAKARA: Daughter of a born slave-maid, have mercy! I won't kill you again! So please save me!

SARVILAKA: You there! Take him away! Noble Charudatta, please order what we should do with this sinful miscreant.

CHARUDATTA: Well, will you do as I tell?

SARVILAKA: What doubt is there about that?

CHARUDATTA: Really?

SARVILAKA: Really.

CHARUDATTA: If so, then let him be quickly—

SARVILAKA: What, slain?

CHARUDATTA: No, no,—*released*!

SARVILAKA: For what reason?

CHARUDATTA:—

If an enemy, having done evil, beseeches protection and throws himself at the feet, he should never be killed with weapons,—

SARVILAKA: Quite so; then he should be devoured by dogs?

CHARUDATTA: No!—

But he should be killed so to say, by being smothered with kindness!

SARVILAKA: Oh Wonderful! What shall I do? Let Your Honour speak further.

CHARUDATTA: So let him go!

SARVILAKA: Well, let him be set free!

SAKARA: Bless me, I am restored to life again! (*exit with the guards*)

(*An uproar offstage*)

(*Again, offstage*)

Here this worthy dame Dhuta, wife of the noble Charudatta, about to enter the blazing fire, even casting off her son who is clinging to her legs and to the hem of her garment, and even though she is being dissuaded by the people whose eyes are brimming with tears!

SARVILAKA (*listening; looking around*): How! Here's Chandanaka!

Well, Chandanaka, what's this?

CHANDANAKA (*entering*): What! Has not Your Honour noticed? There's a big crowd of people to the south of the royal palace. (*he again repeats 'Here this worthy dame Dhuta' et al*) And I said to her: 'Madam, please do not act thus rashly, for the worthy Charudatta is alive.' But when one is affected by grief, who will listen and who will believe?

CHARUDATTA (*in distress*): Oh, my dear, what is it that you have resolved upon, while I am still alive? (*looking upwards* and *heaving a deep sigh*)—

O, you of noble character! Although your deeds are such that the earth is not a fitting residence for them, still, O faithful wife, it is not proper on your part to enjoy the happiness of the next world, leaving your husband behind.

(*He falls in a swoon*)

SARVILAKA: Oh what a mishap!—

We ought to be proceeding there quickly; but here His Honour has swooned! Alas! Our efforts seem to come to naught in all directions!

VASANTASENA: Your Honour should calm yourself! You should go there and induce the worthy lady to live. Otherwise, by such weakness, a disaster will probably befall!

CHARUDATTA (*coming to himself; getting up at once*): O, my dear, where are you? Give me a reply!

CHANDANAKA: Come, sir, come this way!

(*Thereupon all walk about*)

(*Enter* Dhuta *as described, and* Rohasena *clinging to the skirts of her garment, and accompanied by* Vidushaka, *and* Radanika)

DHUTA (*tearfully*): Leave me child! Do not hinder me! I am afraid to live longer lest I hear the evil tidings concerning my husband. (*she gets up; frees her skirt, and walks towards the fire*)

ROHASENA: O Mother, O mother! Please wait for me; for I shall not be able to remain alive without you! (*he swiftly approaches her and again seizes her skirts*)

VIDUSHAKA: As for you, a Brahman lady, the sages have declared it a sin to mount the funeral pyre without the corpse of the husband.

DHUTA: It is better to commit that sin, rather than hear the evil tidings about my husband!

SARVILAKA *(looking before him)*: The good lady is just near the fire; so make haste, make haste!

(Charudatta *hastily walks up*)

DHUTA: O Radanika, please take care of the boy, so that I may do what I wish.

MAID *(mournfully)*: I, too, am going to do what I have learnt from my mistress!

DHUTA *(looking at* Vidushaka): You, sir, please look after the child.

VIDUSHAKA *(with a pang)*: Anybody setting about to accomplish his desire has to follow the lead of a Brahmana. So I will precede Your Ladyship in this matter.

DHUTA: How! They both have refused my request! *(embracing* Rohasena) Child, you yourself should compose yourself that you may offer to both of us the sesamum-mixed water. Of what avail are one's desires, when a thing is past? *(sighing)* For your father will not indeed take care of you!

CHARUDATTA *(hearing it and coming up suddenly)*: I myself will take care of my boy! *(he lifts up the boy with his arms and clasps him to his breast)*

DHUTA *(seeing)*: Wonderful! It seems like the voice of my husband! *(again peering closely; joyfully)* Oh, good luck! It is my lord himself! It is gratifying to me, gratifying indeed!

ROHASENA *(seeing; with delight)*: Wonderful! It is father who is embracing me! *(to* Dhuta) Mother, you are now happy! It is father himself who is taking care of me! *(he embraces him in return)*

CHARUDATTA *(to* Dhuta):—

O my love! What means this terrible attempt on your part while your dearest one still lived? Does the sun-lotus-plant close its eyes, while the sun has not set?

DHUTA: My lord, it is for that that it is described as inanimate. *(seeing; joyfully)* Oh! Wonder! With these same eyes I am beholding my dear friend! Oh, the might of a chaste wife! By merely resolving upon immolating herself in the fire she has been united with her beloved husband! (*to* Charudatta) Victory to my dear friend!

CHARUDATTA: Come, Maitreya! (*embraces him*)

RADANIKA: Oh, the wonderful turn of events! Sir, I salute you! (*she falls at* Charudatta's *feet*)

CHARUDATTA (*laying his hand on her back*): Rise, O Radanika! (*he helps her to rise*)

DHUTA (*seeing* Vasantasena): Good luck! My sister is safe!

VASANTASENA: Now I am safe, indeed!

(*They both embrace each other*)

SARVILAKA: By good luck all Your Honour's friends are now safe!

CHARUDATTA: It is rather by your favour that it is so.

SARVILAKA: Madam Vasantasena, the king is highly pleased with you and has graciously bestowed on you the title of a wedded wife!

VASANTASENA: Sir, my wishes are fulfilled!

SARVILAKA (*pulling the veil* on Vasantasena; *to* Charudatta): Sir, what should be done for this mendicant?

CHARUDATTA: O mendicant, what do you like best?

MENDICANT: Having witnessed this impermanence of things, of such kind, the great liking that I had for an ascetic's life has been intensified!

CHARUDATTA: Friend, his resolution is firm; so let him be made the head of all the Buddhist monasteries in the country!

SARVILAKA: As Your Honour directs.

MENDICANT: It is gratifying to me, gratifying, indeed!

VASANTASENA: Now I feel I have been really restored to life!

SARVILAKA: How is Sthavaraka to be rewarded?

CHARUDATTA: Let that well-behaved servant be freed from his bondage. As for those Chandalas, let them be made the chiefs of all the Chandalas. Let Chandanaka be appointed the head of the police in the country. And, as regards that king's brother-in-law, whatever status he enjoyed formerly, let him continue to enjoy the same even now.

SARVILAKA: It shall be as Your Honour directs. But still, please deliver him over, leave him to me; I will put him to death!

CHARUDATTA: He that seeks protection shall have it! (*he recites the verse beginning 'An enemy who has wrought evil' et al*)

SARVILAKA: Then please say what further pleasure of yours shall be done by me.

CHARUDATTA: Is there any further pleasure beyond this?—

The purity of my character is restored; this enemy of mine, too, who had fallen at my feet, has been allowed to go free. My dear friend Aryaka, who has destroyed his foe from his very roots, is now the king and the ruler of the earth. I have gained this beloved of mine. And you, that are in union with your dear friend, Aryaka, have become my friend. What more, then, remains there to be obtained, that I should now ask of you?—

It empties some and fills others; some it raises up and others it causes to go down; while some it keeps in suspense; thus bringing home to the people this course of the world which consists of opposites mutually meeting together, does Destiny play with all creatures, being bent upon following the maxim of 'the water-wheel and the buckets.'

Still, let there be this—

(*Bharatvakya*[16])

May cows yield plenty of milk, the earth produce all kinds of grains in abundance, the clouds shower rain in due season, and the winds blow so as to gladden the hearts of all people! May all creatures always rejoice, the Brahmanas be ever respected, and the virtuous become rich. And may righteous kings subdue their foes and govern the earth justly!

(*Exit all*)

Here ends Act X, entitled

'The Conclusion'

16. Benediction. See Glossary.

GLOSSARY

Abhinaya: The Sanskrit word *abhinaya* is made of the prefix *abhi* — meaning 'towards' and the root — *ni* — meaning 'to carry'. Thus, it means 'representing (carrying) a play to (towards) spectators'. It is important to note that *acting* is an inadequate translation of *abhinaya* and is an entry point to understand the unambiguous distinction between the Greek (Aristotelian) idea of 'Catharsis' (purgation) and Indian dramaturgical context of 'Rasa' (aesthetic savour).

Bharata's *Natyashastra* enumerates four divisions of *abhinaya*:

a) *Angika*, the artistic gesture

b) *Vaachika*, the modulation of voice and the speech

c) *Aaharya*, the costume and the adornment, &

d) *Saatvika*, the peculiar reactions due to peculiar emotional states.

This fourfold *Abhinaya* performed along with *Sangita* (music), i.e., the triple symphony of *Gita* (song), *Vadya* (instrument) and *Nritya* (dance), produces *Natya* (total theatre), as conceived and described by Bharata.

Abhisarika: Abhisarika is used for a woman who, driven by passion, goes to meet the lover. When a courtesan is referred as *abhisarika*, as a convention, she must go in a splendid dress, with a smiling countenance and with lots of ornaments.

Airavata: The celestial elephant who rose from the ocean when it was churned for *amrit* (ambrosia) at the Beginning; also a reference to the mount of Indra.

Antahpura: Literally 'inner apartment', was traditionally the royal harem of an Indian palace exclusively for women. It is the part of the king's palace where the queen stayed with other court ladies who accompanied her to her bath. Most of the palaces in India have *antahpura* as an integral part. In *Mrichchhakatika*, *antahpura* carries the meaning 'the private space for the inhabitants of the house'.

Ashvaghosa: Ashvaghosa was a distinguished Sanskrit poet who belonged to Saketa (Ayodhya). A Brahman by birth, he later embraced Buddhism. Present in the court of Kushan emperor, Kanishka (first or second century AD), his date is disputed and Western Indologists place him before Kalidasa.

Two works by him, both in the style of *mahakavya*, are extant: the *Buddhacarita* (Life of the Buddha) and the *Saundarananda* (Of Sundari and Nanda). He wrote a play *Sariputtaprakarana* (Sariputta's Prakarana), fragments of which were discovered by Heinrich von Lueders from Turfan in Sinkiang, China and published in 1911.

Asvamedha: One of the significant royal rituals of the Hindu Vedic religion, which finds detailed description in the *Yajurveda*. The Ashvamedha could only be conducted by a king and the objectives were the acquisition of power and glory, gaining sovereignty over neighbouring provinces, and seeking blessings for heirs and the general prosperity of the kingdom. The Ashvamedha celebrated the king as king of the whole world, not just of a part of the world that constituted his kingdom.

Bharata: Bharata or Bharata Muni (sage) is believed to be the creator of the proverbial fifth Veda, *Natyashastra*, the Indian dramaturgical treatise, which also describes him as an austere *rishi* (lit., inspired and learned poet) engaged in penance, pursuit of art, and the wisdom of the *shastras*. As per legend, he had a big family of hundred sons with whom he practised and proliferated the knowledge of drama.

Bharatavakya: The last stanza of the play is referred to as *Bharatavakya*, probably perpetuating the legacy of Bharata, the father of Indian drama. Close to didactic in tone, with an expression of good wishes for general prosperity, it became a convention in classical Indian drama.

Bhasa: Bhasa is the earliest known playwright in Sanskrit theatre, and hence in India. Like most of the Sanskrit playwrights, Bhasa's dates are disputed, but for certain he preceded Kalidasa as the latter reverentially refers to him in his work. If Kalidasa flourished in the first century BC,

Bhasa may be placed in the second or third century BC and if Kalidasa belonged to the fourth century AD, Bhasa cannot be regarded as belonging to a time later than the second century AD.

His plays were discovered by chance in the early years of the twentieth century, among the performance scripts used in the Koodiyattam tradition of Kerala. Ganapati Shastri, a local scholar, attributed thirteen of these texts to Bhasa and published them between 1912 and 1915, though their authorship has been contested. Most of the plays attributed to Bhasa are inspired by the *Ramayana* or *Mahabharata*. While his most popular, *Swapnavasavadatta* (Vasavadatta in the Dream) deals with the popular tale of King Udayana and his queen Vasavadatta, *Urubhanga* (Breaking of the Thigh) a one-act play, departs from the forbidden imperatives of *Natyashastra* by demonstrating the death of Duryodhana on stage. Though the plot belongs to *Mahabharata*, Bhasa made major alterations in *Urubhanga*, the most extreme of which is his portrayal of Duryodhana — whereas in the *Mahabharata*,he is portrayed as a villain, in *Urubhanga*Bhasa gives him more human qualities. While tragedy is rare inSanskrit drama, Bhasa's presentation of Duryodhana's side of the tale adds a certain tragic element to the play.

Daridra Carudatta (The Poor Charudatta) is a fragment in four acts without the initial and the final verses. This play proves to be the germ of Shudraka's *Mrichchhakatika*, a *prakarana* (a play in ten acts). Bhasa's plays form a richly varied corpus both thematically and generically, comprising both full length *natakas* as well as one act plays. Their rich stage potential, in terms of ingenious employment of theatrical techniques, remarkable presence of dramatic irony and humour, powerful representations of emotions, and vast area of human experience, has led to regular productions by modern Indian directors.

Bhava: *Bhava,* or emotion, mood, feeling is the emotive state employed and expressed in performance of a play for the awakening of *rasa*. *Sthayi, Sanchari, Vyabhichari* and *Sattvika* are the four types of *bhavas*, enumerated in *Natyashastra*. To all the *rasas*, there is a corresponding *bhava*.

Bhavabhuti: Born Srikantha at Padmapura in the Vidharbha region of interior Maharashtra, in the seventh-eighth century AD, Bhavabhuti belonged to a respected Brahman family of the Vedic tradition. Considered 'equal to Kalidasa', he became court poet of Kanyakubja, which was later known as Kanauj in present day Uttar Pradesh.

Uttararamacarita (Rama's Later Exploits) is a seven act play that presents the main events of the *Uttarakanda* (Later Book) of the Ramayana. *Malatimadhava* is a *prakarana* in ten acts, a story about the love between Malati and Madhava. Bhavabhuti's works, especially *Uttararamacarita,* offer evidence to his vigorous language, elaborate rather than suggestive descriptions and powerful characterization. Traditionally, Bhavabhuti is compared with Kalidasa with regard to poetic and dramatic excellence, but the former's writing style is distinct and has given predominant space to *vacika*, meaning the verbal aspect.

Brahma: In Hindu mythology, Brahma is the progenitor of the trinity: Brahma, Vishnu and Mahesh. Regarded as the father of Manu and Saraswati, who was also his consort, he is the lord of speech and sound.

Courtesan: *Ganika* is the word used in Sanskrit Literature for courtesan, which may be translated, though inadequately, as an 'elite prostitute'. The *ganika* was a desirable woman, desired for her beauty and as well as her refinement and intellect. Although the ancient Indian texts (*kavyas*) talk about *veshyas* (common prostitutes) and *ganikas* (elite courtesans), they are more eloquent about the latter. In contrast to *veshya*, *ganika* lived in large well-furnished houses as we see in *Mrichchhakatika*, where Vasantasena's mansion has eight grand courts.

As a connoisseur of refined pleasure and culture, the *ganika* was the feminine counterpart of *nagaraka*. Apart from the art of love, the *Kamasutra* gives a long list of arts to be learnt by a *ganika*: etiquette, singing, dancing, playing musical instruments, preparing drinks, performing conjuring tricks, narrating jokes and riddles, staging plays, improvising poetry and knowledge of literature and gambling. The *ganika*'s household establishment was headed by her mother and included maidservants, female messengers, musicians, other professionals and children.

Dasarupaka: *Dasa* meaning 'ten' and *rupaka* as 'drama'. On the basis of subject matter, theme, number of acts *et al*, Indian classical drama is classified into ten major types:

a) *Nataka* (example, *Abhijnanasakuntalam* of Kalidasa)

b) *Prakarana* (example, *Mrichchhakatika* of Shudraka)

c) *Bhana* (example, *Karpuracharita* of Vatsaraja)

d) *Vyayoga* (example, *Madhyamavyoga* of Bhasa)

e) *Samavakara* (example, *Samudramathana* of Vatsraja)

f) *Dima* (example, *Tripuradaha* of Vatsraja)

g) *Ihamriga* (example, *Rukminiharana* of Vatsraja)

h) *Anka* (example, *Sarmishthayayayti*)

i) *Vithi* (example, *Malavika*)

j) *Prahasana* (example, *Mattavilasa* of Mahendravikramavarman)

It is significant to note that the Sanskrit synonym for drama is *rupaka* and not *nataka*.

Dharma: In Monier Williams' *A Sanskrit Dictionary* (1899) the etymological reference of *dharma* is given as 'to hold, bear (also bring forth), carry, maintain, preserve, keep, possess; have, use, employ, practise, undergo'. The classical Sanskrit noun *dharma* is a derivation from the root *dhra*, which means 'to hold, maintain, keep' and takes a meaning of 'what is established or firm', and hence 'law'. *Dharma* is a word without direct translation, but implies 'religion,' 'duty,' and 'righteousness.' The concept behind *dharma* is anything that upholds or sustains a positive order. For instance, an individual, a family, a community, a nation, and the universe all help uphold order. Semantically, it is similar to the Greek idea of *ethos* (fixed decree, statute, law).

Indra: Leader of *devas* (Hindu gods) and *swargalok* (lord of heaven), Indra is the deity of rain and thunderstorm. Wielding *vajra*, the lightening thunderbolt and riding Airavata, the white elephant; at the behest of gods, Indra approaches Brahma to create something which has the qualities of *drashya* (visual) and *shravya* (audible), which led to the genesis of *Natyaveda* (drama theory; *Natyashastra*).

Kalidasa: Author par excellence of *Sringara* (erotic) rasa, Kalidasa is fabled to have been illiterate till he was blessed with wisdom by

Goddess Kali (hence the name 'Kalidasa'). His name has become synonymous with Sanskrit drama. The accounts of his life and times are disputed. For certain, he came after Agnimitra Sunga (2nd century BC) about whom he writes, and before Vatsabhatti (fifth century AD) and Harisena (seventh century AD), authors of two inscriptions mentioning him. Hence the most plausible time frame, according to Indian tradition and modern scholars, may be the first century BC, though Western Indologists place him in the Gupta period, in the fourth century AD. He was the court poet of King Vikramaditya in Ujjain.

Composer of two epics: *Kumarasambhava* (Birth of Kumara) and *Raghuvamsa* (Dynasty of Raghu); two lyrical poems: *Meghaduta* (Cloud Messenger) and *Ritusamhara* (Cycle of Seasons); and three plays *Malavikagnimitra* (Malavika and Agnimitra), *Vikramovarsiya* (Urvasi won by Valour) and *Abhijnanansakuntala* (Sakuntala Recognised), Kalidasa appears to have been deeply imbued with the heritage and travelled throughout the length and breadth of India. His work exhibits comprehensive knowledge of human nature, intense sensitivity to sorrow, love for nature and high idealism.

Kavya: It refers to the literary style in Indian classical literature in the court epics from the early century AD, which is characterised by the abundant usage of figures of speech to create its emotional effects. Other characteristics of the style are hyperbole, the careful use of language to achieve a particular effect, ostentatious display of erudition, and an adroit use of varied and complicated metres, all applied to traditional subjects and themes derived from early popular epics.

The style finds its classical expression in the so called *mahakavya* ("great poem"), in the strophic lyric (a lyric based on a rhythmic system of two or more lines repeated as a unit), and in Sanskrit theatre. The great masters of the *Kavya* form were Ashvaghosa, Kalidasa, Bana, Dandin, Bhavabhuti, and Bharavi. The earliest surviving *Kavya* literature was written by Ashvaghosa, a Buddhist who was present in the court of Kushan emperor, Kanishka (first or second century AD). Two works by him, both in the style of *mahakavya*, are extant: the *Buddhacarita* (Life of the Buddha) and the *Saundarananda* (Of Sundari and Nanda).

Nandi path: A conventional preliminary in Sanskrit drama, translated as 'Benediction'. A poetic invocation to the muse for the success of the play, it is generally performed offstage, followed by the entry of *Sutradhara* (narrator) of the play.

Nataka: One of the ten types of *rupakas* in Sanskrit drama and the most popular one, with five to seven acts. Characterized by the celebrated stories it takes for its plot, the quality of vivacity, portrayal of life on a panoramic canvas, characters contributing to the hero's prosperity, and abundant sentiments of pleasure and pain drawn from the narratives of the *mahakavyas* (epics), historical events and *Puranas. Sringara* (erotic) and *Vira* (heroic) *rasas* are the dominant ones with *adbhuta* (marvellous) rasa arising in the fulfillment at the denouement. The hero would be high-spirited, powerful and virtuous, a god, a demigod, a royal sage, or Kshatriya from a renowned family. He could be *dhirodatta* (firm and noble), *dhiralalita* (firm and playful), *dhiroddhata* (firm and haughty) or *dhiraprashanta* (firm and serene), the popular categories of heroes in Sanskrit drama.

Bhasa's *Swapnavasavadatta* (Vasavadatta in the Dream), Kalidasa's *Abhijnanasakuntalam* (Recognition of Sakuntala) and Bhavabhuti's *Uttararamcarita* (Rama's Later Exploits) are the best known examples of this form.

Natyashastra: Loosely translated as 'a compendium of theatre,' *Natyashastra* is the single most important source to decode and understand the character of classical Indian theatre, poetics, aesthetics, dance and music. An encyclopaedic compendium, it also contains elements of architecture, painting, prosody, language, grammar, phonetics and other aspects of theatre and draws on disciplines as diverse as philosophy, psychology, mythology, ritual and geography.

Natyashastra is a large dramaturgical treatise with about six thousand verses and has been divided into thirty-six chapters detailing all aspects of theory and performance. Drama, an audio-visual form, mirrors all the arts and crafts, higher knowledge, learning, sciences, yoga and conduct. Presently, the only surviving ancient Sanskrit drama theatre tradition is Koodiyattam, which is being preserved in Kerala by the Chakyar community.

For more details, refer to the 'Introduction' and the essays *Origins of Sanskrit Theatre* and *Characteristics of Sanskrit Theatre and Drama* in this edition.

Prakarana: One of the major forms in Sanskrit drama, *Prakarana* etymologically means 'to do perfectly'. In this kind of *rupaka* (drama) of five to ten acts, contrasting with the *nataka*, the dramatist's imagination is more at work as the plot is deeply rooted in the life of common people and social reality. *Sringara* (erotic) and *Prahasana* (comedy) were its principal *rasas*. The hero — a Brahman, minister or merchant of *dhirashanta* types (firm and peaceful) — intends to do meritorious deeds yet has selfish motives, seeking objects of desire and wealth which perish or bring transitory pleasure. The heroine was sometimes a noblewoman and sometimes a *ganika* (courtesan) and often both. *Prakaranas* featured characters like *Vita*, a stock character, the parasitic companion of a prince, of a dissolute young man or of a courtesan, skilled in singing, music and poetry, on familiar terms with his associate whom he serves as a *Vidushaka* (jester). Minor characters included gamblers, thieves and servants. Wider use of Prakrit dialects was done in *Prakaranas*. Shudraka's *Mrichchhakatika* and Bhavabhuti's *Malatimadhava* (Malati and Madhava) are the best examples of this form.

Prastavana: 'Prologue' or 'Prelude' is a preliminary convention in Sanskrit drama which is a part of the first act. The play begins with *Nandi path* (benediction) followed by the entry of *Sutradhar* (narrator) who welcomes the audience and introduces the name and theme of drama to be performed for the day. *Prastavana* (etymologically, the beginning) is interchangeably used with *Sthapna* (literally, to establish) and in case of Bhasa, he only used the word *Sthapna*.

Rasa: The closest definition for the word is 'aesthetic savour'. Translated as 'taste' or 'flavor', *rasa* is the defining feature of classical Indian aesthetic theory. 'Nothing proceeds on the stage without reference to rasa', declares Bharata in *Natyashastra*. Without *bhava* (emotive states), there will be no *rasa*. All the components of the drama beginning with *vibhava* (determinants; stimulus of *bhava*), *anubhava* (consequents, expressive of *bhava*), and *vyabhicharibahava* (the transitory states) collaborate and enable the spectator to experience the *rasa*.

There are nine *rasas* with corresponding *bhavas*. The detailed chapter on rasa theory is prescribed as a background prose reading.

Shiva: The third in the *trimurti* (trinity), Shiva or Mahadeva, also referred as Mahesh, is the destroyer in Hindu mythology, after whom the cult Shaivism is named. Known for both benevolent and fearsome forms, Shiva is regarded as limitless, transcendent, and unchanging.

Sutradhara: *Sutra* meaning 'thread' and *dhar* meaning 'upholder'— a 'thread-holder'. Some scholars argue that the term originated in ancient puppetry in a very literal sense as a 'string-puller'. Sutradhar (narrator) is the first entrant in Sanskrit theatre, who is responsible for auspiciously conducting the *purvaranga* (the preliminaries like Nandi path) before the commencement of the play and delivering the *prastavana* (prologue) and introducing the play. Sutradhar is analogous to a modern director, stage manager and producer. According to Bharata, *sutradhara* must be proficient in playing the various musical instruments, an expert in theatrical speech, gaits and movements of different characters, highly conversant with *rasa* and *bhava*, skilful in the presentation of *natya* (theatre) and other arts and crafts, learned in all the disciplines of knowledge and capable of determining the rhythm and time measure to accompany various songs as well as comprehending the essence of prosody. He should himself be an excellent performer and, after understanding the play, must be able to impart training to other performers.

In the play, he could impersonate a character, but also voiced aloud their inner thought or feelings whenever those were not revealed explicitly. He often intervenes between scenes with metatheatrical remarks about his actors or the action. At the end, he appears again to deliver the *bharatavakya*, the closing stanza of the text, a ritual valedictory blessing.

Ujjayini: One of the earliest human settlements in India, presently known as Ujjain in Madhya Pradesh, Ujjayini was one of the greatest cities in ancient India.

Vidhushaka: This stock character of Sanskrit theatre has invariably influenced all the folk forms and classical traditions of India. The *munshi* of Nautanki, *manasukha* of Raslila, *naqqal* of Nakal, *bhand* of Bandi Pethir, *nanglo* in Bhavai, *songadya* in the Tamasha, *hanumanayaka* or *hasyagana* of the Yakshgana or the *konnagi* of the Kuchipudi, they all owe their existence to *vidhushaka*.

Like *sutradhara*, he also occupies a central position. He creates laughter through instant witticisms and jocular speech, his rotund and ugly appearance, fanciful costume and eccentric gestures. He is a humorous companion and confidante of the hero, and excites mirth by allowing himself to be made the butt of ridicule by almost everybody in a play. In spite of his clownish nature, he is intelligent and therefore the hero's confidant and counsellor, especially in matters of love (for instance Maitreya in Sudrak's *Mrichchhakatika*). He frequently sounds a note of caution to us when he notices something wrong in the hero's behaviour. He has the liberty of commenting satirically on religious institutions, political authorities and social customs. Even in mythological and historical plays, he makes connections to contemporary realities with ease and felicity.

Vishwakarma: The divine architect in Hindu mythology, Vishwakarma is the presiding deity of all artisans and architects. He is believed to be the 'Principal Architect of the Universe', and personifies omnipotence and the abstract form of the creator God (*Vishwa* meaning 'Universe' and *Karma* meaning 'maker' or 'doer').

Yavanika: The ancient Indians generally used 'yavan' for foreigners and more specifically for the Ionian Greeks. *Yavanika* means 'hand-held curtain' in Sanskrit drama. This has encouraged various western scholars to believe in the theory of Greek influence on Indian classical drama, but this is contested and refuted by others.

CRITICISM

Mrichchakatikam: Thematic and Tonal Patterns

Madhusudan Pati

With more than thirty characters to enact the *prakarana* through ten elaborate acts, making use of seven varieties of Prakrit in addition to the chaste Sanşkrit, harnessing almost all the major *rasas*, and pressing into service a wide-ranging social life spanning the plebeian and the aristocratic, Shudraka's *Mrichchakatikam* achieves dramatic synthesis on a large, inclusive scale. For all its vastness of Design, however, there is no sign of structural looseness or diffuseness of dramatic appeal. Selecting an unconventional story of passion and love for its main subject, and inscribing a broader public event of political upheaval into the subsidiary plot, the playwright explores possibilities of integration in respect of various dualities and dichotomies which disturb and dislocate human life at the individual and social levels. The play, therefore, registers the impact of both a great work of art and a profound social document.

More than its large social appeal and the exciting bustle and sweep of theatrical performance, however, the subtle working out of thematic and emotional antinomies in the play should constitute its major interest for us today. Dramatic interest generated by the interaction of characters and development of plot is intensified and expanded here by the submarine engagement of a fleet of thematic 'tensions'. But, as is well-known, Indian critics of the play have, for the most part, concentrated on emotional designing of *rasa* and have neglected the relatively more intellectual area of theme-study. Western critics, on other hand, not suspecting the presence of thematic complexity in Sanskrit plays, have mostly remained content by emphasising the play's realistic appeal. Unlike in the case of many other Sanskrit plays, however, critical concern for thematic structures is of fundamental importance here.[1]

Like all great tragicomedies, *Mrichchakatikam* engages itself in thematic exploration in two main directions: it seeks to discover values which help to sustain a harmonious social order, and to delineate qualities of mind and heart that are necessary for achieving fulfilment at the individual, inter-personal and social levels. It is at once serious and carefree, to treat the comic and tragic potentialities of social life together in order to illumine some higher means to individual fulfilment. As different from a pure tragedy, the social context is constantly kept in view here, and its thematic structures have more of a social than a metaphysical orientation.[2] Similarly, an emphasis on deeper levels of individual being, while envisioning possibilities of human fulfilment, keeps it apart from the traditional comedies. Personal value and social conduct are studied together as means to the establishment of a superior order of harmony and joy. Shudraka's profound dramatic choice of representing a multitude of characters — 'all in the round', as Warder emphasizes[3] — encompasses a wide spectrum of social life. The play, therefore, reveals a significant ordering of dramatic material, comprehending a variety of motifs and themes.

One might, for instance, note the opposition between art and trade which is introduced at many points in the play. On a superficial observation this may appear to be a casual strain, but, in fact, it constitutes one of the fundamental socio-personal themes under examination in the play. Dealing as it does with the sophisticated, metropolitan life in the capital city of Ujjayini, *Mrichchakatikam* attempts a study of the tension generated by values and obligations inhering in two alternative approaches to a vocation — mercenary and aesthetic. It projects the necessity of transforming one's profession into an elevating, artistic experience and thereby deriving a proper sense of fulfilment out of interpersonal relationships. A number of secondary characters in the play enact the experiences of people who seek to bring to their vocations the discrimination and relish of artistic connoisseurs. However ordinary, nay, even debased, be their occupation, the characters display a singular urge to discover therein possibilities of higher refinement and aesthetic gratification. The self-conscious and deliberate manner in which many of them reflect on this theme of 'trade as art' forces one to reexamine certain things in the play which are ordinarily glossed over as incidental, causal matter; and the clear pattern which then emerges, involving the central protagonists as also the secondary characters, convinces one of a conscious dramatic design shaping the theme. Obviously, this is a subject that has particular relevance for the social segment comprising artisans and tradesmen in popular, sophisticated cities. Shudraka's concentration on this theme illuminates one aspect of his *civilised* art.

Even as ordinary a character as Samvahaka, the masseur, is keenly alive to the tension existing between values of art and trade. As he emphasises in course of his talk with Vasantasena in Act II, he used to derive a deep measure of satisfaction in the past by cultivating his delicate art in the service of the noble Charudatta. The rich aesthetic and moral sensibility of Charudatta, Samvahaka indirectly suggests, helped to transform the lives of his attendants. With Charudatta's fortune in decline, however, he no longer has the scope to practise it with elegance and sophistication. Thus there issues a note of deep dissatisfaction as he recounts this story of his life before Vasantasena. That the discontent is a consequence of his present inability to derive any artistic satisfaction from his work is made absolutely clear.[4]

> VASANTASENA : It is a very dainty art that you have learnt.
> SAMVAHAKA : Lady, I learnt it as an art, but it has now become my means of earning a livelihood. (Act II)

Lest the audience should miss the point, the playwright further introduces a pointed comment by an attendant of Vasantasena — 'you have given a reply indicative of extreme dejection' — and underscores the burden of sadness in Samvahaka's voice. Forced to practise his art at a pedestrian level, as merely a trade that brings him his physical upkeep, he suffers an acute psychological depression and goes to the gaming table in search of relief.

A similar urge for aesthetic satisfaction is dramatised even in as uncongenial a situation as house breaking. Shudraka is not bothered by the danger of appearing artificial and fastidious as he presents Sarvilaka soliloquising on the scientific skill and artistic elegance he wished to display in digging a passage through the wall of Charudatta's house. It is remarkable that Sarvilaka is allowed to adopt an almost poetic attitude in such an inapposite context. A note of artistic delight is unmistakeable in his words:[5]

> Should I dig the hole in the shape of a full-blown lotus or a half-moon, or a spacious pond, a triangle or a swastik or a full pitcher? Where and how should I demonstrate my art so that the townsfolk tomorrow would be wonderstruck by looking at it? (Act III)

In a later scene Chardudatta is presented observing the breach and admiring the beauty and precision of the digging. Charudatta's words on that occasion not only reveal his own extraordinary sensitiveness to beauty at all times, but also vindicate the uncommon skill of an artistic temperament of the burglar.

It may be noted here that Sarvilaka is not just an ordinary thief. He is, no doubt, adept in house-breaking, but he looks upon his skill as an artistic accomplishment and a thrilling game, and not as a means of livelihood. His aspiration and taste are seen to belong to a much higher order. His present exploit, for instance, is a desperate venture to acquire a huge ransom to liberate the girl he loves from her bonded state. And, even so, he expresses a moral revulsion against the deed. The later scene bring to the fore an altogether different dimension of his character — his capacity for noble, unselfish actions. As he is on his way to consummate his union with his beloved, he comes to know that his 'friend' Aryaka was being persecuted by the king and was in desperate straits. At once he resolves to do his duty, and turning his back upon the felicity he had been so assiduously pursuing, rushes to the aid of the political rebel. A similar sensitiveness and capacity for lofty actions distinguishes Samvahaka, too. The passion with which he reacts to life's chequered shade, and turns his back upon worldly allurements to join an order of monks, shows only the reverse side of a coin whose obverse face had already been presented in the beginning in his singular reactions to the quality of his trade.

Both Sarvilaka and Samvahaka dramatise the appropriation of artistic values to commonplace — even abject — states of life, and reveal in their different ways the process of enriching individual and social existence. Equally, they convey the message that unless the professional self receives a heightening touch and gains an aesthetic dimension, it may ultimately lead to a state of emotional atrophy and moral debasement. Enrichment of life through the adoption of an artistically and morally elevating perspective, and degeneration of life in the absence of it — this may, in fact, be one way of summing up the entire subject-matter of the play. Shudraka dramatises the interaction between values of trade and art in such an inclusive thematic context.

Thus we find a remarkable interplay between the professional wooer and sensitive connoisseur, between the mind of an artful pander and that of an artist, in the character of the Vita. As a love-companion to Shakara, the king's brother-in-law, he is first of all seen performing his duty of wooing the reluctant courtesan, Vasantasena, with wonderful wit and felicity. The sensitiveness and refinement reflected in his speech clearly indicate an essentially poetic disposition.[6] Very soon we notice how his artistic temperament (a qualification that is prescribed by tradition for such a character) makes him despise certain actions he is called upon to perform as a professional wooer. Being employed by an uncouth, lustful character is a matter of profound dissatisfaction to him. While he performs his duty of furthering Shakara's amorous interests, displaying superb elegance and lyrical verve, he also experiences deep psychological discomfort over this aesthetically and morally repellent task of aiding brutish lust. Thus, in the course of his very first appearance itself, while he reels off a series of brilliant poetic arguments, enticing Vasantasena to his master, he also expresses his disgust at Shakara's boorishness in a number of brief asides.

Once again, we have here a case similar to Samavahaka's 'trade minus art'. When it caters to physical passion, his service declines into a trade; but, in the context of the higher romantic love, it grows into an exquisite instrument of aesthetic gratification. It is the artist in him who feels revolted by the incongruity implicit in the picture of Shakara as a love of Vasantasena. Therefore, even while campaigning so brilliantly in favour of his master in Act I, he experiences a sense of personal satisfaction when he comes to know that she was spurning all enticements because of her deep love for Charudatta, and helps her escape from the clutches of Shakara. This he does with no hope of personal gain; indeed, his professional prospects the other way should be far more alluring. He comes to her aid only because Vasantasena's union with Charudatta is an aesthetically satisfying proposition: it would be a 'fitting union of two gems',[7] as he says. The play of moral and aesthetic sentiments is thus a constant feature in the character of the Vita.

He is pulled in two contrary directions all along: his sense of duty and material prospects keep him tethered to Shakara, but his moral and aesthetic susceptibilities incline him towards the noble and indigent Charudatta. It is a pull between the demands of trade and art. There is, of course, not the slightest doubt in the mind of the spectator as to which of the two forces would gain predominance, but the dramatic excitement over the confrontation of values is kept alive by the manner in which Shudraka introduces it time and again in the character of the Vita, and presents it against both comic and tragic backgrounds in the play. The pleasure-garden scene in Act VIII furnishes the finest illustration of this collision of loyalties in the mind of the Vita.

The scene opens with Vasantasena boarding a wrong carriage and, instead of the rendezvous fixed with Charudatta, reaching the spot where Shakara is diverting himself with his companions. The Vita is the first one to peep inside and discover her in the carriage. In shocked disbelief he stares at her for a few moments, thinking she was on a visitation to Shakara. He reflects ruefully on the gross incompatibility of the match,[8] but does not immediately voice his inner feelings. As a professional love-messenger he is somewhat cynical about women's love,[9] and, after all, in his role of an attendant to Shakara he can have no justification for expressing a feeling of disappointment. All the same, the demands of 'art' and 'trade' are in a state of acute conflict within him and all his pronouncements and gestures in this memorable scene are informed by the resultant tension.

To begin with, he softly mutters his disapproval of Vasantasena's conduct, telling her that she showed a lack of 'nobility' and consistency in first fleeing from Shakara and then returning to woo the same person, under pressure of her mother, for the sake of financial gain. Immediately, however, the professional in him takes over, and affably, though cynically, he comments:[10]

> Why, I did say to you:
> 'O good maiden, serve equally one much coveted by you and one disliked by you. (Act VIII)

But no sooner does she protest her innocence and claims she is there by accident and not by choice, he is anxious to whisk her away to safety. He fails in his attempt, no doubt; but the alacrity and zeal with which he volunteers to help her, even by running a grave personal risk, go to suggest an exceptional temperament in the Vita. Clearly, both in his disparaging asides with regard to Shakara and his joyous envisioning of the union of Ujjayni's *'nagara Sri' with 'nagarayām* tilakabhutam'[11], his reactions are as much moral as aesthetic. One should compare the reactions of the Cheta and the Vita to the main events in the scene to appreciate the special quality about the latter. The Cheta talks in broadly human terms, and the sense of outrage he expresses on the strangling of Vasantasena is a moral reaction, but they have nothing of the special aesthetic dimension so prominent in the words[12] and acts of the Vita.

As the scene slowly advances to his horrible climax, the Vita is shown gradually shedding his professional mask and revealing his true identity. The manner in which he champions Vasantasena's attachment to Charudatta, expresses his disapproval of her suspected change, laments on her supposed death, and, finally, refuses to serve the hateful Shakara and chooses instead to join the rebel army, clearly indicate that an essentially artistic temperament impels him to perform these gallant, self-abnegating acts. To be sure, he keeps up his professional posture almost till the very end. For example, he offers resistance when Shakara tries to force Vasantasena, but when Shakara pleads with him to leave them alone for a few moments so that he might try to win her over with tender solicitation, the Vita sees he cannot really protest, and with a certain professional cynicism walks away some distance. But this final act of spurning Shakara and running away to the rebel camp convincingly demonstrates the triumph of his artistic disposition over his professional, mercenary self. He has, all along, made a valiant effort to subserve to the demands of his trade without compromising his own personal set of values. But when 'trade' demands a total disregard of human values and emotions that he had nurtured with an artist's passion, when it compels decline into grossness, he chooses rather to sacrifice the trade itself and court a life of danger and insecurity.

King Palaka is not brought on the stage. But we are made to know of his vileness from what we hear in his proclamations and orders, and from the conduct of his protege, the barbarous Shakara. Ranged against this established order of primitive sensuality and gross material affluence, we have the opposite camp constituting a refined order of innocence, beauty, and tender sentiments. The uprising in the play does not have merely a political dimension; it also vindicates a refined order of art and morality against a gross order of art and morality, against a gross order of material advancement and sensual appetite. The play is concerned with a reorganisation of social order and individual life on the basis of certain aesthetic, liberal sentiments that facilitates fulfilment of joyous potentialities of life both at communal and personal levels. It is for this reason that the artistic disposition of most of the characters, primary and secondary, has been so often emphasized. Rebhila the musician (whom we do not see), Vasantasena and Charudatta the lovers, all the secondary characters except Shakara, and, finally, Aryaka, the new king, represent the ascendency of artistic sentiments over mundane interests and pulls. The motif attains a splendid consummation in the story of Vasantasena.

Vasantasena is one of the most captivating of the heroines in Sanskrit drama. As her name suggests, and as the play poetically underscores at serval points, she represents the enchanting possibilities of spring. She trades in love, but is no harlot; a sophisticated *ganika*, she is coveted for her culture, vivacity, good taste, elegance, and artistic accomplishments, and not merely for her extraordinary physical charms. She heads a huge establishment with à large retinue of attendants, musicians and receptionists, and has a mansion which with its seven courtyards and sprawling enclosures, as described vividly in Act IV, vie in opulence with a royal palace. To all appearances, therefore, she has everything that life has to offer. But, in reality, she is a deeply discontented soul, pining for a life of fuller womanhood. The art of love in which she is adept, and her artistic and feminine attainments have so far been instruments of merely material advancement. She now longs to convert them into instruments of a profounder emotional fulfilment. She is in quest of a love which would transform her life, ensuring social and moral elevation at the same time as it would satisfy all her feminine longings and enrich her sensibility.

Charudatta embodies exciting possibilities of such a fulfilment. He is a high-souled noble of Ujjayini, noted for his moral excellence, generosity of spirit and artistic temperament, qualities that have resulted in his decline from a high state of prosperity to one of mortifying penury. But the reversal of fortune has in no way affected his righteousness and grace. Vasantasena cherishes a deep love for him, but has not had any opportunity of communicating her feelings. When the play begins, she is already a love-lorn woman, spurning all enticements of wealth and honour from other quarters given over entirely to found thoughts of Charudatta. It so happens that, one evening, seeking to escape from the clutches of Shakara who is in hot pursuit, she runs into the old mansion of Charudatta. She has then her first opportunity of observing him at close quarters and engaging him in a talk. Conducting herself in a demure and polite manner she exploits this opportunity for a shrewd assessment of his feelings for her. Charudatta, who had some prior intimation on Vasantasena's love for him, is quite eager to respond, but feels inhibited because of his present indigence. With unerring insight Vasantasena gauges his mind and, prudently enough, decides to wait for some more time before revealing herself fully to him. She knows it would be necessary, first of all, to convince him of the genuineness of her passion and to make him aware of the nature of her longing. And, for all her infatuation, she would not force matters. Deftly creating an opportunity for a further meeting with him, therefore, she happily returns home.

A number of interesting vignettes are offered in course of the first four acts to bring out the nature of Vasantasena's ardour and aspiration. Through wonderfully subtle and elegant strokes, the fuller dimensions of her passion for Charudatta are slowly unfolded before the audience. It is important to note here how, in his dramatisation of the possibilities of human happiness, Shudraka does not introduce a set of ideal values to the exclusion of immediate, earthly ones. In contravention of conventional puritan codes, the play tries to portray how sensual charm itself can be harnessed as a refined source of fulfilment. Vasantasena's passion emphatically includes a longing for amorous joys, but it is transformed into something far more fascinating by being manipulated only as a stage in the process that would ultimately lead to her final liberation — both socially and psychologically — from the trade into which she is born. For her, love is an instrument of emancipation from a state of emotional deprivation and moral inferiority. She aspires to be not just a beloved, but a wedded wife of Charudatta's, and looks upon the emotion of love as the means to a profoundly satisfying human relationship, and not mere sexual gratification. Moreover, as the 'clay-cart' episode in Act VI so movingly dramatises,[13] she also wishes her passion to consummate in the fuller glory of motherhood.

She is keenly conscious of the irony and incongruity of her longing in the context of the realities of her life. This poignant awareness is movingly conveyed through her soliloquies and dialogues in the first four acts. Thus three more acts pass before she has a second meeting with Charudatta. A supremely tender sequence is employed in the interval to unfold the drama of Vasantasena's inner hopes and fears in the privacy of her house. Her struggle against the obligations of her station in life, her search for unadulterated love, her quest for the true beauty of life against the background of a sordid set of mercenary, decadent values which threaten to keep her permanently enthralled, constitute the main subject of *Mrichchakatikam*. In this context, Shakara and Charudatta represent two opposed worlds: the world of trade which debases life, and the world of art which refines and elevates it. Keeping the personality of Vasantasena at the centre, the playwright offers a series of developing tensions between the two worlds in order to demonstrate how love as an art can be an aid to the highest kind of fulfilment and operate as a liberating principle of life when its base and mercenary interests are purged away. The excellence of Charudatta's personality and his indigence, together, are so often emphasized by the dramatist in order that the nature of Vasantasena's choice — '*Ata eva Kāmyate*' (declared twice, in Act II and V) — may be properly appreciated.

When this is understood to be the subject of the play, certain apparently ordinary elements take on a highly significant dramatic value. Since Vasantasena's love for Charudatta is divested of all ornamental and mercenary trappings, the first four acts which gradually lead up to the union of the lovers are carefully designed to highlight the process in a variety of ways. Vasantasena's apprehensions over the prejudice of her 'art' would naturally evoke in others by being associated with 'trade' is repeatedly suggested through a number of poignant situations and dialogues. More interestingly, dramatic intimations are furnished at four crucial stages of Vasantasena's relationship with Charudatta to emphasise the necessity for her to discard all artful enticements and appurtenance natural to her trade.

On the eve of her very first meeting with Chārudatta the situation so develops that she has to remove her garland and anklets before she can enter his house.[14] Later, as she prepares to go on a visitation to Charudatta, she is shown rejecting all artful devices counselled by her attendant and going forward to express her love in a simple, spontaneous manner. Just before her union with Charudatta takes place, similarly, she has to remove her rich fineries and drape around herself a piece of ordinary cloth.[15] In the last scene of the play, again, when her reunion with Charudatta is dramatised[16], and when a permanent relationship with him is legitimised by the king's orders, she is presented in a plain, unadorned state. Dhuta's ringing statement that her husband is her brightest ornament,[17] and that she needs no other embellishment, similarly underscores this opposition in the play between plainness and adornment. In fact, the dramatic justification for conveying the story of union after Act V through a series of melodramatic situations spanning another five acts resides mainly in the fact that the playwright is engaged in demonstrating how love which liberates and blesses is one where, in place of the 'abhisarika'— image and sportive disposition presented in Act V and the beginning of Act VI, a sober, unostentatious, sedate approach to life, as projected through the last scene is more appropriate and enriching.

The two climactic points in the play, at the fifth and tenth Act, are designed to project two contrasted sets of values. One of these involves an approach to human passion that is at once moral and aesthetic, and the other is concerned with divesting love of all elegant make-up. Together, they indicate the manner in which joys of human life acquire fullness and mellowness. From art it is necessary to glean sentiment, elegance, refinement and aesthetic impulse. But if art is to serve a richer human life, and not to degenerate into mere trade, it would be necessary to peel off all its embellishment and artificiality which are sanctioned by society on grounds of refinement and sophistication. Spontaneity and sophistication are both necessary; but spontaneity, not of the primitive kind, and sophistication, not of the affected sort. Thus the play dramatises a healthy fusion of art and life, and thereby shows how to enrich either.

One remembers many of the delightful touches that were introduced by the playwright so casually and unostentatiously in the earlier scenes: Vasantasena smelling the jasmine-scented cloak of Charudatta and happily commenting to herself that he was still not averse to the pleasures of youth; Vasantasena dissuading Samvahaka from taking a precipitate decision to enter the holy order; Charudatta joyously recounting before Maitreya the enthralling charms of Rebhila's music; and a host of others. One also remembers scenes containing a more direct and extended celebration of the joys of life and references to the experience of anguish in their absence — in the middle acts and the final act, in particular — to realize how the playwright is not out to persuade the audience to spurn either riches, or sophistication, or amorous delights, as such. Even in his darkest days of alienation and indigence Charudatta continues to nurture his aesthetic sensibility. Vasantasena, similarly, is presented as a woman with deep, overpowering impulses, and also as a supremely elegant person possessed of a highly refined artistic taste. The contrasted figure of Shakara highlights the theme in the reverse, and shows how, life of brute passion, devoid of the aesthetic dimension, grows more and more debased, repellent and sinister.

There are yet other thematic opposites organised into the plot and the characterization in this play. We have, for example, an interesting set of ironies and contrasts produced by an interplay of the themes of affluence and indigence. In the first half the obvious contrast is the one between Charudatta's decrepit mansion and Vasantasena's splendid palace. These two constitute the setting of the drama for its first five acts during which the dramatist repeatedly refers to the magnificence and splendour of the one, and the dilapidated state of the other. But the emphasis is also ironical: in terms of moral and emotional quality, the poor state of Charudatta's household symbolizes enviable prosperity, and the splendid establishment of Vasantasena represents deprivation and degeneration. But we are not offered a simplistic equation of poverty and virtue, and of riches and vice, for if affluence can be dangerous and suicidal as in the case of Shakara, poverty, too, can be demoralising and crippling, as in the case of Sarvilaka. If affluence is suffocating to Vasantasena, indigence is depressing and agonising to Charudatta. As in the case of other dualities, here, too, Shudraka orders the resolution only after portraying an interaction between the two sets of values over an extended length of the play.

Attention has already been drawn to the suggestive manner in which values of simplicity and plainness have been projected through certain important situations in Vasantasena's life. A dramatic climax relating to the thematic confrontation between affluence and indigence is also organised with similar finesse and emotional appeal at the very centre of the play. Soon after Vasantasena's final and unequivocal rejection of 'trade' and acceptance of an elevating relationship of love is dramatised at the end of Act V, Shudraka introduces the illuminating clay-cart episode in the beginning of Act VI to drive home the message from yet another perspective. Not only is Vasantasena's gift spurned there by Dhuta (with the polite, but telling, disclaimer: 'my husband is my most precious ornament'), even the child Rohasena is shown instinctively recoiling from her generous presents and demonstrative affection. This is a dramatic manner of emphasising the fact that Vasantasena has not yet attained the simplicity and natural dignity of Dhuta: her acceptance in that household, therefore, is still very provisional. Rohasena reacts, first, at her 'costly jewellery', and, next, at her lack of composure, and protests that she does not at all resemble his mother. While this interlude enhances the sympathy of the audience for Vasantasena by a moving representation of her yearning for motherhood, it also *places* her. Rohasena's reactions suggest, at one level, Dhuta's reaction as well — it is by no means a hostile response, but not a cordial one either. The message is clear. Vasantasena can be admitted into their affections — that is to say, she can properly fulfil her aspiration — only on the basis of pure, natural, and noble impulses, as in the last act, and not of glamour and sentiment as at present.

That, incidentally, gives us the cue to the real significance of the title[18] *The Clay-Cart*. A golden cart, the play seems to suggest, need not be frowned upon if it serves to support a refined order of life and does not contribute to a vulgarisation and debasement of society. There is nothing of the puritan about the writer of *Mrichchakatikam*, but there is nothing of the dandy either. While he raises no objections against the golden cart as such, he does not hesitate to suggest that one is more likely to find a safer means of conveyance to the felicities of life in a plain and simple clay-cart. What the golden cart stands for may be welcome to the extent that it would make possible the leisured cultivation of arts and virtues, as indicated through references to Charudatta's earlier life. But, then, one must guard against frivolity and vulgar complacency which a leisured life may breed, and assiduously cultivate certain higher values and impulses. As the Vita comments, a pond with hygienic, sweet water is likely to dry up soon (like Charudatta), but is to be preferred far more to a stale and contaminated one where the water goes on accumulating to no end[19] (as with Shakara or with Vasantasena's mother in the beginning). Ultimately, the value of a thing has to be determined by its viability as a means to a richer, more satisfying life. Shudraka is not a prudish, reactionary mind — he is most not opposed to sex and riches in themselves; he welcomes them as things which make life more enjoyable, and rejects them only where they debase it.

As the mansions of Charudatta and Vasantasena constitute the background for the first six acts, the pleasure-garden and execution-ground are the major settings for all but one of the later acts. These settings, too, involve a fundamental irony: the pleasure-garden is the scene of brutish violence and murder, whereas the execution-ground is the place for the enactment of the story of emancipation and regeneration. As the tension between wealth and penury in the initial acts is sorted out at aesthetic and emotional levels, so also in the final acts a complementary duality and resolution is worked out. Instead of the contrast between Charudatta and Shakara that was earlier foregrounded in that context, now the theme is amplified through the contrast between Charudatta and Samvahaka.

With a sudden change of fortune, Charudatta regains in the last scene his lost riches and authority, and also his lady-love. He accepts the return of good fortune with humility but expresses his sense of joy in clear, unambiguous terms. Samvahaka the masseur, on the other hand, whose feeling of discontent had made him embrace the life of a Buddhist monk, reaffirms in the same scene his conviction that fulfilment of life can never be attained through worldly prosperities, and firmly rejects the authority and wealth offered to him. He is thus at the other extreme from Shakara, for whom sensual joys and riches constitute the highest felicity. Between the two extremes of vulgar affluence and ascetic renunciation, Charudatta and Vasantasena provide a mature, refined human approach to the wealth-penury opposition.[20]

The Samvahaka-Charudatta contrast highlights yet another tension: worldliness versus other-worldliness. This is rather a surprising dramatic opposition to have in a play which so insistently and ardently advocates joys of human life. Both in style and theme so predominantly mundane, it yet introduces certain elements to set off the opposed values of renunciation and other-worldliness. Shudraka's catholicism of taste and inclusiveness of perspective are reflected in his refusal to project the bliss of union achieved by Charudatta and Vasantasena as the only worthy vision of human happiness. He does not, to be sure, repudiate it with the bitter cynicism of a Malvolio or a Jacques. But he faithfully notes that a third vision is still possible, beyond the apparently exhaustive dualities of good and evil, comedy and poetry, trade and art, affluence and poverty, which he has brought into play and resolved with a bias for the wholesome joys of human life. He brings in, therefore, the Buddhistic vision of the world as a wheel of suffering, indicates the Vedantic direction of gaining absolute serenity and bliss, and also adumbrates a fatalistic vision of human life with Destiny controlling the entire chairoscuro.

At the very beginning of the play, in the benedictory invocation itself, the opposition between love and renunciation, passion and void, life in *Prakriti* and life of the *Purusha,* are juxtaposed and fused by presenting, first, the picture of Lord Shiva in *nirvikalpa samadhi,* absorbed in the Great Void, and next, the contrasted picture of the same Lord Shiva in an amorous posture with His consort Gouri, the embodiment of *Prakriti.* The first involves total detachment, a state of complete self-sufficiency; the second shows a state of reaching out, of joyous union. If the second is intended to provide a key to the main *shringara rasa* of the play, the first would naturally suggest the highest form of *santa* in which ultimately everything finds its rest and fulfilment, and to which, according to Indian philosophy, all other experiences of life should serve as a stage and a process. Similarly, in the penultimate verse of the play — the one conveying Charudatta's philosophical reflections on destiny — a number of the earlier references to fate are gathered together and sounded to a final crescendo, counterbalancing the sense of fulfilment and joy issuing through the lover's union at the personal level, and the, political reorganisation at the communal level. Both reversal and consummation are viewed as being equally inevitable, and equally temporary, underscoring a philosophical vision of life's varieties.[21]

> It empties some and fills others; some it raises up and others it causes to go down; while some it keeps in suspense; thus bringing

> home to the people this course of the world which consists of opposites mutually meeting together, does Destiny play with all creatures, being bent upon following the maxim of 'the water-wheel and the buckets.
>
> (Act X)

As has already been stated, confrontation between values of sensuous enjoyment and Buddhistic renunciation is portrayed in the character of Samvahaka both at the initial and concluding stages of the play. What seemed on the earlier occasion a desperate, precipitate resolution comes off in the last act as an attainment of philosophical maturity. The masseur who wanted to grow into a mendicant friar in Act II, we remember, had been solicitously cautioned by Vasantasena not to throw away, on a sudden impulse, all the sensuous delights of human life. While her words on that occasion revealed her own pirations and values in a superbly controlled, dramatic manner, they also served to bring into focus the Great Crossing. Now as the play draws to a close, while a state of harmony and joy is shown descending all around, Samvahaka is invited to state his choice:

> CHARUDATTA: O mendicant, what do you like best?
> MENDICANT: Having witnessed this impermanence of things, of such kind, the great liking that I had for an ascetic's life has been intensified!
>
> (Act X)

The very things that yield for the central protagonists a feeling of deep satisfaction, and fill their minds with exciting future possibilities, induce in the monk an intenser realisation of their inconstancy and convince him that Buddhistic *nivritti* is the only answer to the problems life. The playwright does not forcefully argue it out; he does not wish to negate the vision of human happiness he has carved out with the story of the union of Charudatta and Vasantasena — a vision, it may remembered which embraces sensuous, moral, and aesthetic values, and suggests a state of peace and harmony at the individual, domestic, and communal levels. No doubt, it is an exciting and glorious vision. Yet, in the interest of comprehensiveness of portrayal, Shudraka feels impelled to organise a brief but telling vignette, emphasising certain values which transcend the purely humanistic ones projected through most part of the play.

That *Mrichchakatikam* represents the high watermark of realism on the Sanskrit stage needs no emphasis. This has already received an abundance of critical comments and Western critics, in particular, have found in it something very germane, and have rated the play very high on that score. It is necessary to point out, however, that Shudraka's realism is not confined to etching out life-like scenes and furnishing superb examples of dramatic conversation. More importantly, it involves the projection of a special aesthetic position and, moral perspective. It is necessary, again, to emphasise that an urge for realism in the ordinary sense does not control the portrayal of life here. There might be a number of scenes which are preeminently realistic in treatment and inspiration. But scenes presenting life in an idealised, heightened style are still more numerous, and they constitute the heart of the play. What one actually witnesses is an extended interplay between realism and idealism, both in terms of impulse and treatment, running through the entire play.

In exploring superior possibilities of human happiness, Shudraka brings into play a continuous interaction between values of pragmatism and idealism. He has, of course, incorporated a wide variety of realistic characters and situations, thus exhibiting a penchant for realism in Chaucerian or Shakespearean manner. But, at the same time, the central episodes in the play have been imbued with moral, emotional, and aesthetic heightening of such an order that it cannot just be described as a realistic piece. Vasantasena's passion, Charudatta's personality, Dhutta's character, nay, even the actions and aspirations of secondary characters like Sarvilaka, Madanika and Samvahaka, are pitched on such an idealised plane that *realism* can hardly be a suitable descriptive term for the dramatic portrayal. On the other hand, the idealisation is not of a fictive kind, introducing incredible action, or heroic sentiment and conduct. Even the melodrama in the final act cannot in that sense be dismissed as unrealistic. Shudraka's effort at synthesizing the real and the ideal in envisioning possibilities of happiness in human life involves a corresponding fusion at the presentational and stylistic levels.

In presenting an integrated vision of the comic and the poetic, the playwright devises a recurrent pattern of joining together gay and solemn tones. In the dialogues of the characters, no matter which pair it is, one notices a continuous interplay of light banter and serious concern. Except for the two melodramatic scenes, this integration of the light and the serious is noticeable all through the play, imparting to the development of action and character a superb thickness of dramatic texture. How pervasive and exciting this interplay is may be easily appreciated by consulting not only major scenes like the meeting between Charudatta and Vasantasena, but even such apparently casual dialogues as are exchanged between Vasantasena and her attendants in Acts II and IV. What we have is not merely a mechanical alternation of comic and serious scenes involving secondary and central characters respectively, as seen in many other plays; in *Mrichchakatikam,* the casual and solemn operate concurrently nearly all the time. For example, even so palpable and typical a comic figure as Maitreya has to wear his comic gusto, for the most part, as a shield against the sad and serious thoughts rising within him with overpowering force.

At the stylistic level, similarly, one finds Shudraka constantly harnessing the dual modes of documentation and symbolisation, prose and poetry. Fusion of statement and metaphor results in an exciting dramatic communication. At the conclusion of Act I, for instance, we hear a verse recited by Charudatta,[22] which constitutes both a realistic description of landscape and a brilliant projection of mood and impulse. Following the use of a significant pun on '*sneha*' (meaning both *oil* and *affection),* words like 'mud' and 'milk' acquire symbolic value and build a delightful network of emotive and semantic suggestions. Similarly complex evocations are incorporated at a number of other places. One needs only to remember the use made of a host of things — a shawl, moonlight, necklace, storm, clay-cart, and many others — to realise the dramatic power with which realistic delineation and symbolic projection have been integrated in the play.

The introduction itself provides clear indications of the tonal design. The solemnity and poetic fervour of the benedictory verses very soon give place to an irreverent exclamation of the stage-manager: 'Enough of this verbiage'. The comment is, no doubt, conventional. But, as may be seen, it sets the tone for the entire introductory scene. After making this remark the stage-manager himself grows rather high-serious and gives a long discourse. This, in turn, is followed by a comic interlude, something very racy, earthy, and completely uninhibited and direct. Thus we find a certain repetition of a tonal pattern — solemn formality is first deflated through down-to-earth realism, then there is the recurrence of a serious mood which, in its turn, is punctured by a light, informal touch. This alternation of formal and informal styles, of solemn and earthy moods in the Induction, prepares the ground for a larger and far more elaborate manipulation of alternating, opposed tones and values in the play proper. The stage-manager's resolve to switch over from formal, cultivated Sanskrit to a more easy and suitable colloquialism is thus highly significant. Apart from some very fundamental critical and historical questions it might stimulate in respect of a possible reformation of dramatic style, or expansion of audience, that might have occurred in the time of Shudraka, the immediate dramatic interest of this unusual introduction is clear: it suggests the playwright's intention of emphasising both an opposition and integration of certain values and perspectives.

It is necessary to remember that Shudraka is neither rejecting nor adhering exclusively to any one of the two modes or values. His vision of human happiness is built upon a synthesis of the two approaches to life: pragmatic and aesthetic, realistic and idealistic, informal and sophisticated. And so he juxtaposes and balances both the connected styles and perspectives. Thus, for instance, Charudatta's aesthetic, aristocratic outlook on life is counterbalanced by the prudent, matter-of-fact approach of Maitreya who carries into the play the stage-manager's iconoclasm, his intolerance of poetic sentiments and his empiricism.

We see sudden delightful incursions of Maitreya's earthly humour into situations that are highly charged with sentiment, deflating the pressure, but not quite destroying the poetry[23] of it. Thus, while Charudatta and Vasantasena in the first act bow to each other, highly-strung and uneasy with the passion and embarrassment of first encounter, Maitreya solemnly mimicks their gesture and comments, he is "like a half-grown camel bobbing his head between two verdant patches of rich corn, with their stalks bent under the wind".[24] One may see how his words effect dispersal of tension without spoiling the sentimental beauty of the occasion. The half-grown camel is, of course, brought in for sheer mirth, offering a grotesque caricature of the stiff and formal manner in which the lovers are bowing to each other. But there is also the other, counterbalancing image of pure beauty. The poetic evocation contained in the graceful wavy movement of ricestalks — symbolising elegance, youthfulness and fertility — intensifies rather than disturbs the romantic mood. Maitreya's words beautifully unite the comic and the poetic and evoke both laughter and thrill in a chaste, elegant manner. Similarly, again, towards the end of Act V when emotional fervour in the lovers has attained its peak-intensity, and the approaching high-romantic consummation is prepared for with a compelling lyricism, Maitreya's mock-innocent questions on the purpose of Vasantasena's visit and her plans for the night introduce a sharp anti-climactic earthliness into a rarefied, idealised situation. And here, too, comedy does not obstruct the movement into a deeper lyrical experience; rather it imparts a touch of reality, and enhances the appeal.

It is instructive to note how this endeavour at harnessing various dualities in respect of a number of themes controls the manipulation of both style and vision in the play. Contrasted elements such as personal and communal, contrived and accidental (for example, the progress of Vasantasena's ornaments in the early acts as also at Act IX) humanistic and fatalistic, are joined as closely fitting arcs to prepare the comic-poetic world of the play. More because of such thematic complexity and multivalence than that of organisation of a number of episodes ('pataka' of the political revolution, and a number of *prakaris* relating to a variety of minor events), a vast gallery of dramatis personae, numerous vividly contrasted settings and situations, and the sprawling expanse of ten acts, that *Mrichchakatikam* produces its impact of vast, versatile sweep and comprehensive portrayal. In the pursuit of a truly satisfying life, a life that is joyous, wholesome, and noble, it explores a vast and variegated world full of dualities and contrasts. Without forcing those opposites and alternatives into some arbitrary, contrived mould, it attempts to synthesize as many of them as possible and, for the rest, leaves the options open.

The play certainly upholds the Classical approach of the golden mean, but does not look askance at man's devotion to extreme values as long as that does not affect the sweetness and harmony of social life. Everything is welcomed: everything except evil and ugliness. As the verses at the two climactic points in the play — concluding ones in the supremely graceful fifth act and the superbly melodramatic tenth act — equally emphasize, the notes may be different, and the experiences may vary, but what ultimately emerges from their orchestration is an exquisite harmony, bringing about an unforgettable sense of joyous fulfilment all around:[25]

> Shrilly on the *Ida* leaves, rumbling on the branches, harshly on stones and fiercely on the waters — thus do the showers fall, resembling lutes played upon in musical concerts to the keeping of time. (Act V)

> May cows yield (plenty of) milk, the earth produce all kinds of grains in abundance, the clouds shower rains in due season, and the winds blow so as to gladden the hearts of all people! May all creatures always rejoice, the Brahmanas be ever respected, and the virtuous become rich. And may righteous kings subdue their foes and govern the earth. (Act X)

Notes

1. References to the text are from *The Mrichchakatikam,* M.R. Kale (ed.), Motilal Banrasidass, Varanasi, Third Edition, 1972. Sudraka's efforts at striking a balance between certain opposed themes have been noted. Thus, for instance, A.K. Warder (*Indian Kavya Literature,* Vol III, Motilal Banrasidass, Delhi, First Edition, 1977 — Section 1183 and 1211) talks of the theme of fate and free-will. But the range and multiplicity of the syntheses have not been adequately discussed yet. As Raymond Williams notes (*Modern Tragedy,* London, 1966, p.24), transition from a metaphysical to a critical orientation is a natural development in the sphere of tragedy proper, too. But the centrality of the individual in a tragedy differentiates it from a tragicomedy like *Mrichchakatikam* which is both individual and general in its orientation at the same time.
2. A. K. Warder, op. Cit, 1211.
3. *Mrichchakatikam,* Act II.
4. Ibid., III. 13.
5. For example, see, Theodore Goldstucker, *Sanskrit and Culture,* Indological Book House, Varanasi, 1971, p.18.

   How erroneous it is to describe him as a "parasite" can be seen from the present discussion. Sylvain Levi is so impressed by this character as to eulogise him in very extraordinary terms and, in the process, commit a grave critical error:

   "The Vita is the most important character of all; we should almost say the only intelligent character of the Indian theatre".

6. Sylvain Levi, *The Theatre of India,* Vol.I, Writer's Workshop, Calcutta, 1978, p. 186.
7. *Mrichchakatikam,* Act I: " Ratnam Ratnena Sangachchate" (A Jewel unites with a jewel - Tr. Kale, op. Cit., p. 39).
8. Ibid. VIII. 16: (Seeing Vasantasena — sorrowfully to himself) — Ah! How is it that the doe is following a tiger! Oh, alas! —
9. Ibid., VIII. 17 — "Asaundiryam".
10. Ibid., I.31 ("Samupachara bhadre supriyam Vapriyam Va") " Embodiment of the prosperity and beauty of the city"' and "the auspicious vermillion mark at the centre of the forehead".
11. *Mrichchakatikam,* VIII 38:

    (Coming to himself, mournfully). O Vasantasena! The river of the water in the form of innosense is dried up. Love has departed to its original home! Alas, alas! O you who were an adornment of your ornaments, you with that sweet face, . . . you that were the river of goodness, with laughter for the sandy bank, . . . Alas! Alas! . . . .

    (Tr. Kale, op. Cit., p. 293)
12. Ibid, Act VI.
13. Ibid., Act I.
14. Ibid., Act V.
15. Ibid., Act X.
16. Ibid., Act VI — (Aryaputra eva mamabharana-visesa yanatu bhavan).
17. An extreme example may be cited.
18. Srinivasa Sastri discusses the significance of the title and, in line with almost all other commentators of mrichchakatikam, first of all, emphasises the importance — in terms of development and plot — of the jewels left in the clay-cart by Vasantasena. He quotes from Kantanath Sastri's *Mrichchakatikam Samikhya* and argues that the clay-cart signifies dangers of discontent and jealousy. This is clearly an untenable argument. One notes in this context his strange assertion that Charudatta was drawn towards Vasantasena because of his dissatisfaction with Dhuta, and that this caused all the misery — hence, he claims, the title suggests the feelings of discontent. There could hardly be a more exotic interpretation. See, Dr. Srinivas Sastri (ed.), *Mrichchakatikam,* Sahitya Bhandar, Meerut, Fourth Impression, 1976, pp. 16-7. Dr. Rangeya Raghav goes to another extreme by bringing in (in its clay-cart, i.e. physical body) to explain the title.

19. *Mrichchakatikam*, I. 46:

"He has dried up by satisfying the thirst of men, like a reservoir full of water, in hot season." (Tr. Kale op. Cit. P.51)

20. Critics who do not enquire into the deeper thematic implications of the play unjustly condemn the repetition of the motif, and some other central elements in the play:

"It does not seem proper to decry poverty through nearly fifty-two verses and to discuss it at forty places (in the play). Is penury so reprehensible a thing? . . . Personally, I would like to suggest that his (Charudatta's) scornful attitude against the state of indidence is in itself the cause of his fall from the exalted state of Brahmanhood. How wide, indeed, is the gulf between the state of a Brahmin and the joys of the brothel-house!"

Ramji Upadhyaya and Ram Gopal Misra, *Sanskrit Ke Mahakavi Aur Kavya*, Ramnarayallal Venimadhab, Allahabad, 1965, p. 117.

The reiteration of a certain theme should have alerted these critics to look for some deeper significance. The second half of the passage betrays not only a fundamental misconception in respect of the personalities of Charudatta and Vasantasena, it also shows an inadequate critical interest in the traditional prescriptions relating to a mixed kind of *Prakarana* like *Mrichchakatikam*. Also the narrow, prudish approach is fundamentally at variance with the spirit and culture implicit in the play. Charudatta is nowhere represented as a profligate or a lascivious person. The superb balance in Sudraka's representation has to be appreciated if one is to properly understand the play. As their further arguments reveal, the authors do not also correctly respond to dramatic stylisation.

Similarly, Bhola Shankar Vyas (*Sanskrita Kavi Darsana*, Chowkhamba Vidhyabhavan, Varanasi, 1968 — p. 290 & 296) reacts unjustly against Charudatta's laments on poverty, and against political stylisation at Act V.

21. *Mrichchakatikam,* X. 60 (M.R. Kale, op. Cit. P. 407)

22. Ibid, I. 57.

23. G. K. Bhat makes a neat assessment of Maitreya's clear sightedness and matter-of-factness:

"Maitreya turns his sharp satire to every inconsistency and shortcoming Nothing seems to escape his observan and thoughtful eye. He does not exclude Charudatta or even himself from criticism. It is no wonder then that the entire creation figures in his witty similes. From the clod of earth to the heaven, from flowers to stars, from the inanimate creation to the animals and human beings, and from zoology to mythology, all things swim into his unchecked vision and supply the theme for his sharp and satirical humour." — G. K. Bhat, *The Vidusaka*, The New Order Book Co., Ahmedabad, 1959, p. 237.

But the fact that Maitreya is also intrinsically in sympathy with poetic sentiments and higher ideals of life has to be adequately understood. Shudraka would not present a one-sided Vidusaka in a play *Mrichchakatikam* which is a marvel of synthesis and balance.

24. *Mrichchakatikam*, I, 56.
25. *Ibid.*, Kale op. cit.,p. 213 and 407.

Mrichchhakatika: A Reading

Devendra Raj Ankur

It is often understood that the first four acts of Sudraka's *Mrichchhakatika* have been lifted verbatim off Bhasa's *Charudatta*. It is correct that *Charudatta* indeed is the fundamental source of inspiration for the writing of *Mrichchhakatika*, but if we study the first four acts of both these plays, we can clearly see many differences between the two. Undoubtedly, the sequence of events in the first four acts is the same in both these plays but the resemblance ends right there. A comparative analysis of the two plays from the prologue till the end of Act IV reveals some facts and truths about both. Why is Bhasa's creation considered to be incomplete, even though at the end of the fourth act, Vasantasena is seen getting ready with her friend for her abhisara[1] with Charudatta? It seems that the love story Bhasa intends to represent through the characters of Charudatta and Vasantasena comes to a conclusive climax by the end of Act IV. However, if the original play is complete, then why does Sudraka feel the need to take the story further? Even if we consider the play to be indeed incomplete, the question remains as to what was missing in the story that inspired Sudraka to compose *Mrichchhakatika*.

In my opinion, there are a number of points, *in toto*, in Bhasa's *Charudatta*, which could not have gone unnoticed from Sudraka's eye. Love between a Brahmana merchant and a ganika[2] is rare in Sanskrit dramatic literature. The love contexts of *nagarvadhus* like Amrapali and Chitralekha from ancient Indian history, folklore and mythology have always attracted authors, filmmakers, readers and spectators. The single most important reason for their popularity is that they represent a certain novelty and modernism in their treatment of love. Sudraka would have clearly realized the exploratory possibilities of a number of characters from *Charudatta*; for instance — characters like Sakara (King Palaka's brother-in-law); a degraded and corrupt king like Palaka; Samvahaka, the gambler, who never reappears after Act II in

Charudatta. If King Palaka is destined to die, then doesn't the invention of a character like Aryaka remind us of Sudraka's own story? Aryaka and Sudraka are very similar names, and both had devastated the unrestrained power of the establishment.

Naturally, Sudraka wouldn't have found a better opportunity to explore and articulate this unutilized narrative of potential in the earlier text, than in *Mrichchhakatika*, and he indeed exploits it optimally. Let's consider the prologue. The inaugural stage direction in the *prastavana* (prologue) itself reveals a differentiated beginning. Even in Bhasa, the *Sutradhara* (narrator) enters after the *Nandi* (benediction) but the *Nandi* is not performed on stage. Hence the implication is that the narrator enters after the *Nandi* is performed. But in *Mrichchhakatika*, the *Nandi path* is performed on-stage, and who is reciting it — the Sutradhara or the chorus? If it is the Sutradhara, then why is the Sutradhara's 'Entry' mentioned as a stage direction? If it is the chorus, then the 'Entry' of the narrator can be justified. Sudraka might be following the *Natyashastra* traditions, according to which, the Sthapaka's entry happens after the Sutradhara's exit. Needless to repeat, often all playwrights have preferred using 'Sutradhara' over 'Sthapaka'. Even Bhasa employs the term 'Sutradhara', even though he uses 'Sthapana' instead of 'Prastavana'. In Sutradhara's first dialogue itself, Sudraka displays his dramatic skill: "Enough of an exertion that causes distraction in the curiosity of the audience."

A comment such as this aimed at the audience evokes laughter and creates a blissful ambience which is necessary for relishing the 'rasa' of this play. It is here that we identify *Mrichchhakatika* as a *Prakarana*. It must be noted that the term 'Nataka' is not used for *Mrichchhakatika*. This clarifies that there is a distinction between a Prakarana and a Nataka.

Mrichchhakatika's prologue is the only prologue in Sanskrit dramatic literature that presents a detailed account of its composer's life and his death. Apart from this, the subject-matter of the play is presented with similar elaboration. After reading these two descriptions of the author and the text, it appears that Sudraka is drawing our attention to the representation rather than the story itself.

Sudraka has also painted Bhasa's Sutradhara in a new light. Sudraka's Sutradhara is not simply the Nati's husband but he is a trained singer and musician in his own right. *Charudatta's* Sutradhara starves unexplained, whereas Sudraka represents his unbearable hunger as his devotion for sangeet[3]. Like the entry, the exits of the two Sutradharas are different as well. Bhasa's Sutradhara exits after inviting Maitreya but the Sutradhara in *Mrichchhakatika* leaves on his quest for another Brahmana after Maitreya's refusal. *Charudatta*'s prologue does not clarify the location of the play nor the characters among whom the action unfolds. However, Sudraka has depicted the flourishing town of Avantipuri or Ujjayini, where finding even one indigent Brahmana is rare.

Two other significant changes in Sudraka's prologue need attention. Firstly, he does not use either 'Sthapana' or 'Prastavana' to describe the prologue. Instead, it is designated as 'aamukh' (Preamble in English). The second difference is regarding the dialogue style of *Mrichchhakatika*. This difference is in the use of Prakrit in a Brahmana's (Sutradhara) conversations with his wife, the *Nati*. It is perhaps the first instance in Sanskrit drama where the male protagonist, and at that, a male Brahmana converses with a female character in Prakrit. Is it because *Mrichchhakatika* is a *Prakarana* and such liberty is permitted there? Or is it because the play proliferates in thieves, gamblers, masseurs, prostitutes, etc. — all earthy characters — that a glimpse into its language is made available in the Sutradhara's use of Prakrit?

In this way, we see that Sudraka, with his unique imagination, constructs the prologue differently from that of Bhasa, and hints that even as he retains every word of the first four acts of Bhasa, the final warps and wefts of his play make the spectator/reader forget *Charudatta* completely. That *Mrichchhakatika* is his only surviving and available play allows us to get a taste of his dramatic prowess. To be sure, we find *Padmaprabhritkam*, a *Bhana*[4] by Sudraka, but this solo *rupaka* has neither the space for a prologue nor any scope for narrative and imaginative innovation. This is the reason why *Mrichchhakatika* stands out in the repertoire of Sanskrit drama and it shall always remain as contemporary and modern as it would have been in its own time.

These were a few points related to the prologue. To proceed to the first four acts of *Mrichchhakatika* that are based on Bhasa's *Charudatta*: Sudraka's impulse for innovation and newness is clearly visible in these four acts. For instance, in the first act of Bhasa's *Charudatta*, when Sakara enters the stage chasing Vasantasena, he is accompanied only by one more character who we know as Vita, but Sudraka adds another character, Cheta, to make Sakara's entry more impressive. These characters help in the elaboration of various narrative tropes that appear after the first four acts.

Cheta regularly intervenes in acts six, eight and ten, creating for himself a unique identity which can hardly be anticipated in the first act. The same can be said for Vita. In the eighth act, Vita is present with Sakara before and even after Vasantasena's arrival. In this way, Bhasa's Vita, whose association with Sakara is only a formality and compulsion, takes the courageous decision of leaving Sakara, in the eighth act of Sudraka's play.

Sudraka has completely changed the structure of the second act of Bhasa's *Charudatta*. In *Charudatta*, the Samvahaka enters in a straightforward and blunt manner but Sudraka's visual imagination of Samvahaka is incredible from the point of view of dramaturgy. First, Samvahaka being chased by Dyutkara and Mathura, him hiding in a Sun Temple behind the goddess' idol, Dyutkara and Mathura's entry in search for Samvahaka, them gambling before the idol and Samvahaka's eventual dashing into the gambling — all these are Sudraka's fundamental inventions that provide to the actor innumerable possibilities of *angik abhinaya*[5] while also generating laughter and dramatic energy in the narrative of the play. After this entire context, can we still conclude that Sudraka has copied Bhasa's play verbatim?

Although the third act of Sudraka does not vary much from Bhasa's in terms of the sequence of events, the delineation of Sarvilaka's character far surpasses Bhasa's Sajjlaka. Sarvilaka is depicted as a *sah-nayak*[6], an alter ego to Charudatta — both characters are Brahmanas, are lovers of *ganika*, and leave no stone unturned to fulfil their love. The only difference is that while it is Vasantasena who makes the first move in the former case, it is Sarvilaka who does so in the latter. Sarvilaka decides to steal in order to emancipate Madanika from Vasantasena's servitude. Though even stealing is considered to be one of the sixty-four arts in the Indian tradition, there is no instance of its use in any play prior to *Mrichchhakatika*. After this, one encounters it in *Prabuddharohanaya* in the medieval period, and more recently, in *Charandas Chor*. It requires no reiteration that *Mrichchhakatika* is an inspiration to both these plays. Moreover, another important detail becomes conspicuous with regard to Sarvilaka here: after freeing Madanika from Vasantasena, Sarvilaka also brings her the social acceptance of a *vadhu*[7], in the next act itself. The fourth act of *Mrichchhakatika* deviates significantly from Bhasa's play. The plot takes an unpredictable turn just as Sarvilaka brings Madanika out of Vasantasena's brothel. We are informed from the backstage that Aryaka, who was imprisoned by King Palaka, has fled and announcements are being made to arrest him. This difficult moment of choice that stares him in the face — on the one hand the newly-wed bride and on the other, his friend Aryaka, who is on the go to fight the uncurbed, despotic and corrupt political regime — has been sensitively etched out by Sudraka. Leaving his wife at merchant Rebhila's place (a friend of Charudatta), Sarvilaka joins the imminent mass movement against King Palaka. This subplot is a significant point of departure which has no basis in Bhasa's *Charudatta*. Sudraka has already introduced this subplot in Act II when Darduraka, after his quarrel with Dyutkara and Mathura, runs to support the revolt which is consolidating to usurp the political regime.

In this very act, Sudraka's portrait of Vasantasena's large mansion opens the actor to the multiple possibilities of *angik abhinaya*. However, it is equally true that all the productions of *Mrichchhakatika* in the last 30-40 years, with the sole exception of Habib Tanvir's, have omitted this scene. One wonders as to how this scene must have been enacted in the first production of *Mrichchhakatika*. There is not the slightest doubt that an actor can bring this scene to life. A similar scene is found in Kalidasa's *Vikramovarshiyam*, where Urvashi transforms into a vine in the thick forest while Pururava keeps trying to find her. Coincidentally, in both the plays, this scene appears in the fourth act. Presumably, there could not have been much realism in stage setting in those days. The actors would have presented the whole scene entirely through *angik* and *vachik abhinaya*. It is possible that Kalidasa drew inspiration for one complete act of his play, from Vasantasena's eight-storeyed large mansion in the fourth act of *Mrichchhakatika*.

It is possibly the first time in the history of Sanskrit drama, indeed Indian drama, that a play has separate titles for each act. For instance, Act I is called 'Depositing of the Ornament', Act II 'Samvahaka, the Gambler', Act III 'The Making of a Breach', Act IV 'Madanika and Sarvilaka', Act V 'A Stormy Day', Act VI 'The Exchange of the Carriages', Act VII 'The Escape of Aryaka', Act VIII 'The Killing of Vasantasena', Act IX 'The Trial', and Act X 'The Conclusion'. The only other instance of separate titles for each act is the plays of Bertolt Brecht, the twentieth century German playwright. Before each episode of his play, he would also tell the gist of the narrative to the spectators through painted curtains and slides. In *Mrichchhakatika*, the title is placed at the end of the act but the possibility of placing it in the beginning exists too — through posters, slides and strips. It would not be surprising if this technique was used in those days.

The fifth act of the play again highlights a deeper meaning. The same Vasantasena, who decorates Madanika as a *vadhu* in the fourth act, gets the right to spend the night in the *antahpura*[8] of Charudatta's home. This right, in a manner, implies the social acceptance of the Charudatta and Vasantasena relationship, including that by Charudatta's wife Dhuta.

On the subject of ornaments, it is interesting to note how Sudraka has employed a dramatic tactic. Ornaments are constantly at play in every act. In Act One, Vasantasena entrusts all her ornaments in a bowl to Charudatta. In Act Two, Vasantasena's gold bangle helps free Samvahaka from the debt of gambling. Sarvilaka steals the same bowl of Vasantasena's ornaments in Act Three which in Act Four he submits to Vasantasena herself to free Madanika. Also, in this act, the *ratnavali* sent by Dhuta becomes a part of the ornaments. In the fifth act, Vasantasena adorns these ornaments again when in torrential rains she goes to Charudatta for *abhisara*. In Act Six, Vasantasena fills Charudatta's son's toy clay-cart with these ornaments. In Acts Six and Seven, the iron shackles on Aryaka's foot become symbols of Vasantasena's ornaments for the cart-driver. In Act Eight, the Mendicant, Samvahaka recognizes the unconscious Vasantasena through the ornaments in her hand and aids her immediately. In Act Nine, the ornaments which Charudatta consigns Vidushaka to return to Vasantasena (she had loaded them on Rohasena's toy cart in Act Six), fall off from the bundle. Lastly in Act Ten, again we encounter these ornaments through Cheta, given to him by Sakara in order to not disclose the secret. In this way, the constant presence of the ornaments through all the acts becomes a dramatic ploy that induces new turns in the narrative each time. The use of a ring, in Kalidasa's *Abhijnanashakuntalam* is another example of a similar dramatic strategy. In the same vein, in Vishakhadatta's *Mudrarakshasa*, the ring of Rakshasa (the last minister of Nanda dynasty in ancient India) causes the turbulence in the plot.

In the sixth act, similar to the ornaments trope, Sudraka uses another dramatic tactic, unparalleled in the world dramatic literature, and that is the mistaken swapping of carts — Aryaka in Charudutta's cart and Vasantasena in Sakara's, and the chaos that ensues. In a way, the next four acts revolve around this particular event. Though not identical, a similar strategy is used in Shakespeare's *Comedy of Errors* where a confusion in the identities of the identical twins leads to absurd situations. In the Italian playwright C. Goldoni's *The Servant of Two Masters*, similar comic situations unfold as the narrative progresses. Even in Hindi films, such ploys have been repeatedly used and for this, they owe tribute to *Mrichchhakatika*. The *Natyashastra* scholar C. Bryski's argument about the influence of Sanskrit plays on Hindi films seems appropriate and *Mrichchhakatika* has probably contributed the greatest towards this. The swapping of carts is also an elaboration of the toy cart of Charudatta's son, Rohasena, which kept him restless.

Let us discuss two situations in Act Six before proceeding further. The first is the entry of Charudatta's son, Rohasena. Sanskrit plays often have child characters: Duryodhan's son, Durjaya in Bhasa's *Urubhanga*; Dushyant's son Bharata in Kalidasa's *Abhijnana-shakuntalam*; Rakshasa's son in Vishakhadatta's *Mudrarakshasa*; and Rama's sons Luv and Kush in Bhavbhuti's *Uttaramcharita* — these characters make the densest and most complex situations poignant. In *Mrichchhakatika*, Rohasena appears in Acts Six and Ten, when Charudatta is about to be hanged. Seeing his father in such a situation, Rohasena's poignant outpouring transforms the *Sringara Rasa*[9] that dominates the play to *Karuna Rasa*[10]. The second situation arises when the state police officers, Viraka and Chandanaka check the cart in which Aryaka is riding. Their bickering reminds us of the two guards in *Andhayug*[11], and Anuswar and Anunasik of *Ashadh Ka Ek Din*[12]. These dramatists often use such dramatic techniques to present diverging points of view. Needless to say, these moments infuse laughter into the play. In the short story genre, Premchand's *Shatranj Ke Khiladi* and Bhagwati Charan Verma's *Do Baanke* would present similar examples. One cannot forget Mohan Rakesh's Ram Bharose and Shyam Bharose in his short play *Bohot Bada Sawaal*. We can easily infer that the tradition of comic situations was pioneered by *Mrichchhakatika* centuries ago.

It is often said about Sanskrit drama that it does not adhere to unities of time, space and action, unlike Greek Drama. However, if we study *Mrichchhakatika* in this context, we observe that the total timeframe of action in this ten act play is limited to four or five days. The first act encompasses the events of one day from morning to night, the second act begins with the morning of the second day and the third act starts at midnight after the second day and continues till the next dawn. Act IV follows the same morning to evening sequence. Act V concerns the same evening. The first five acts, as such, end within the first three days of the play. Similarly, the events from Act VI to Act X are divided over the timeframe of the next two days. For example, Act VI commences with the morning of the fourth day, Act VII that same afternoon and Act VIII ends in the evening, which means three acts thread the time from dawn to dusk of one day. Act IX begins with the morning of the fifth day and finally the tenth act commences with the same afternoon, and progresses towards the end in the evening. One can also probe the time-span of these ten acts from another perspective. Almost every act is framed between morning and evening, with the only exception of the fifth act, in which Vasantasena moves out of her mansion in the deep dark of the night for her *abhisara* with Charudatta. The unity of space can be analyzed similarly. In each act, action is centered at a particular space; notwithstanding the fact that the action of the play continues in the *antrang* (indoor) and *bahirang* (outdoor) spaces simultaneously.

Mrichchhkatika's courtroom scene in Act IX is truly incredible. Today we read about corruption, bribery, murder, rape, theft, robbery in the newspaper — does this courtroom scene seem any different? Perhaps this is one of the reasons why a play like *Mrichchhakatika*, whenever staged, seems to be as contemporary and timeless as it must have been in its own time. Maybe the character of law is such that in whatever age it is represented, it will always be similar — a perennially petrified Judge, a Sakara who pressurizes officers and an innocent Charudatta's increasing isolation — do we ever feel that we are reading an almost two thousand year old text?

Even as Bharata may have prohibited scenes of lovemaking, eating, death etc., in the *Natyashastra*, the playwright of *Mrichchhakatika* renders these norms irrelevant. In this regard, Sudraka's imagination of the scene where Charudatta is heading towards his execution is unmatched. Charudatta *gheraoed* by two *chandalas*, being taken all over the town, announcing his crime at various spots — all this adds a different dramatic intensity to the play.

The reference to the *vadhsthala*[13] appears in *Mudrarakshasha* as well but it is too brief and without the long journey before death. The same can be said about Bharatendu's *Andher Nagari*. If one has to refer to a Western play, then the German playwright Karl Georg Büchner's *Danton's Death* can be remembered. It cannot be denied that a hanging scene in any play or film, besides being pitiable and poignant, is also theatrical and in a play like *Mrichchhakatika*, its presence seems to be inevitable.

For the first time, we witness a play where all the characters except Sakara belong to the middle class or the lower class. Charudatta may have been a powerful businessman of Ujjayini once but his present circumstances are no different from the other characters. On the other hand, Vasantasena, however much wealth and riches she may have under her possession, is no more than a whore or *nagarvadhu* in the social hierarchy, even as she is respected as an artist, being an adept dancer. Here, we also remember Sartre's play *Respectable Prostitute*. It is difficult to find such a multifaceted portrayal of the 'underworld' of the society in India's dramatic corpus. One can invoke *Andher Nagari* and *Agra Bazaar* here but in these two plays also, the characters do not return once they leave, whereas in *Mrichchhakatika* they repeatedly meet, collide, separate and thus stimulate a unique pace in the play.

Plots containing conspiracies against corrupt states and their eventual dethroning have been recurrent. The credit for inaugurating this trend should completely go to *Mrichchhakatika*. The revolution against the state can be seen in Büchner's *Danton's Death* as well but it doesn't spring from the lower strata of society; in *Mrichchhakatika*, each and every character of the play is involved in the people's revolution. At times, the Charudatta-Vasantasena love story seems to recede to the background, while the action involving Aryaka, Sarvilaka, Darduraka, Samvahaka and others rises strongly to the fore. Is it a mere coincidence that a king is not the protagonist in either *Mrichchhakatika* or *Mudrarakshasa*? Even if the king is present as the context, the objective of the plays is to usurp him. The two plays are also similar in that they pioneer the tradition of political drama — *Mrichchhakatika* followed by *Mudrarakshasa*.

The trend of symbolic titles started with Bhasa's *Swapanvasavdatta*. On the other hand, it is equally true that the title of Sudraka's main source of *Mrichchhakatika* — *Charudatta* — is simple and straightforward. Just as Sudraka nuances and expands the plot of his play, he conceptualizes a completely new title such that it bears no resemblance to *Charudatta*. By naming it *Mrichchhakatika*, did he want to establish that though the central idea of the play is taken from *Charudatta*, he is certainly composing an original play? This is why he takes inspiration for the title from Bhasa's second play, *Swapanvasavdatta*. Subsequent plays like Kalidasa's *Abhijnanashakuntalam*, Vishakhadatta's *Mudrarakshasa* and Bhavbhuti's *Uttarramcharita* continue this tradition of symbolic titles. It is difficult to determine what Sudraka conceived first — the title of the play or the context of Charudatta's son with a clay cart in Act VI. Not just this, there are other recurring appearances of the cart from Act VI to VIII. There is no mention of the cart in the first five acts except the brief reference in Act II where Vasantasena's attendant informs her of the rich merchant's cart awaiting her. Hence, it is established on its own that the scenes involving Charudatta's son and his cart, and the subsequent scenes of Charudatta's and Sakara's bullock carts, are Sudraka's own inventions. The question still remains as to what a title like *Mrichchhakatika* or *Little Clay Cart* suggests. What is this clay cart and what does it symbolize? Does it symbolize Charudatta's fall to penury or does it refer to the wealthy Vasantasena's decision to renounce her wealth to live with an impoverished Charudatta? Would it much rather be linked to Rohasena's clay toy and his demand for a golden one? Does it refer to the quick succession of events around the swapping of carts in the latter half of the play? Should it be seen as referring to the arrest of Aryaka by King Palaka and his subsequent resurgence with Sarvilaka's support in the end? Lastly, apart from these textual references, should the cart be read as a metaphor for the complex journey of life in which innumerable colors, forms and lines are intertwined?

At the level of structure, *Mrichchhakatika* represents an example of a *Prakarana*, as enumerated in the *dasarupaka*[14] in the *Natyashastra*. According to the *Natyashastra*, the plot of a *Prakarana* depends on the imagination of the writer. Sringara is its chief rasa, its vritti Kaushiki,[15] and it is mostly a love story of a Brahmana hero and a *ganika*. Till date, there is no other manuscript of a complete *Prakarana* available. As such, *Mrichchhakatika* is considered to be the only example of this form. Another feature of *Mrichchhakatika* is that while it is a *Prakarana*, it encompasses the other nine categories of the *dasarupaka*. The symptoms of so many types of drama are not found in any other Sanskrit play. The protagonist Charudatta is a *dhiralalit* hero[16] who is not only fond of music, but is possibly a fine singer and musician himself. This is evident from the conversation between Vidushaka Maitreya and Charudatta on their return from Rebhila's musical recital as well as through Sarvilaka's description of the musical instruments at Charudatta's house. It is after watching Vasantasena's dance performance that he feels affection for her. This confirms his inclination towards the arts. In this way, *Mrichchhkatika* runs along two parallels: the erotic and delicate love story of Charudatta and Vasantasena, and second, a strand of the social reality of the time emerging through thieves, gamblers and corrupt officers. Both these parallels frequently collide with each other and this interaction gives birth to incidents which take the storyline in umpteen directions.

To sum up, in the nearly one thousand years old Sanskrit drama tradition, Sudraka's *Mrichchhakatika* holds a unique position. *Mrichchhakatika* or *The Little Clay Cart*, on the one hand is a love story of Charudatta and Vasantasena, on the other hand, the play paints a powerful and vivid picture of political, social and religious circumstances that naturally establishes this play as an incredibly contemporary text. The characters of the thief, gambler, brother-in-law of the king, masseur and corrupt officer, coupled with a mass revolution which succeeds in usurping the political establishment, together give *Mrichchhakatika* a character which is rarely found in the entire corpus of Indian dramatic literature.

Notes

1. Rendezvous of lovers.
2. Courtesan. See Glossary.
3. Refers to *gayan, vadya, nritya*; not just music.
4. Another category of drama, enumerated by the *Natyashastra*.
5. Physical or corporeal drama.
6. Supporting hero.

7. Pertaining to "bride", but not literally.
8. Interior quadrangle. See Glossary.
9. Erotic sentiment.
10. Compassion sentiment.
11. A play by Dharamvir Bharati.
12. A play by Mohan Rakesh.
13. Execution house.
14. Ten kinds of plays, as categorized in the *Natyashastra*. See Glossary.
15. A kind of dramatic representation focusing on dance, songs and luxury of all kinds.
16. A conventional hero of a Sanskrit play who is polite and good-natured. and a lover of art forms, like Udayan of *Swapnavasavdatta*.

Devendra Raj Ankur in conversation with Shantanu Bose

Shantanu: Why has Sanskrit Drama been made a part of the curriculum of the National School of Drama?

Devendra Raj Ankur: The most important objective is to acquaint the students with a variety of dramas, styles and forms in the three year curriculum. Also, we want to see how the present generation constructively negotiates and responds to the twenty five hundred year old dramatic tradition. If we can identify more with plays of foreign origin like *Oedipus, Antigone, Medea* etc., then trying to find meaning in the dramas of our own tradition becomes more valuable. A theoretical framework when put to practice offers impossible propositions but along with this, one needs to understand, it brings along possibilities of creating anew. It is quite a challenge for an actor to perform Sanskrit Drama because it is in itself an ensemble of alienation, connection, speech, story-telling etc., which are considered to be essential elements for an actor today. An actor therefore gets an opportunity to imbibe all these essentials.

Shantanu: Why *Mrichchhkatikam*?

Devendra Raj Ankur: I had performed a few Sanskrit plays like *Swapnavasavadatta, Mattavilasa, Shakantulam, Urubhang, Madhyamvyayog,* but not *Mrichchhkatikam*. This was one of the reasons I chose to take it up. Also, there is no other play in Sanskrit dramatic literature as relevant to contemporary times as *Mrichchhkatikam*. The names of the characters may be archaic, but events are relevant. Unlike in other Sanskrit plays, the characters and events here are not unilateral. Moreover, the equation between Dhuta and Vasantasena does not reappear between a wife and a beloved in any other play. A society with a variety of relationships and restrictions, a socially discarded woman and the courageous act of reclaiming social prestige and acceptance to her — these are scenarios pertinent even today. The present system is so deeply corrupt that it is unable to bring any change in the life of a common man; people have to fight their battles all by themselves. Around us, there are so many Vasantasenas and Shakars alive today. *Mitti Ki Gaadi*, therefore, represents the effort of the common man to rise above his situation.

Shantanu: How has it been to work with young actors?

Devendra Raj Ankur: The association of young actors with theatre is itself profound. They can play a child, an old man, all they have to do is to free themselves from preconceived notions. We believe in the process and hope our young actors are also endeavouring to adopt it.

Shantanu: Any concluding remarks?

Devendra Raj Ankur: This is a collaborative performance by the team. The student actors and directors have shaped it from the beginning till the end. With no prior association with any specific dramatic tradition, they have evolved their own new style which is a blend of folk, classical and contemporary. The environment in which this play reached its fruition has been open and receptive.

Translated from Hindi

[Interview culled from the brochure of Sudraka's *Mrichchhakatika*, performed by the second year students of National School of Drama, New Delhi from November 19-22, 1993 at Sriram Centre, Mandi House, New Delhi.]

From Page to Stage :
A Photo-essay on *Mrichchhakatika*

Kuljeet Singh

Enormous joy is in watching the *written* word, *perform*.
Although there is a sublime contentment in articulating an idea as a written word, which one would have visualized in the nascent process of developing a script for the drama, nothing can replace the pleasure to see it performed.
A play is a *play* is a *play*.

No writer has ever written a piece of drama to be taught in the classroom. As students of literature, when we endeavour to understand the same piece of writing in the classroom, it often ends up giving us a limited sense of the performing text and more so when the text is as complex as Sudraka's *Mrichchhakatika*. A clichéd proverb vis-à-vis 'image', yet with a lot of wisdom, 'is worth a thousand words' stands true for, at least, theatre.

This photo-essay attempts to put together a few photographs from the production of *Mrichchhakatika* by the students of National School of Drama, New Delhi, directed by KS Rajendran in 2008, to create a visual journey. Nothing can replace witnessing the play but the intent of this essay is to break the routine of having only the texual analysis and attempt to glimpse through and understand the stagecraft, semiotics, and the challenges of staging a Sanskrit drama in contemporary paraphernalia.

Photographs Courtesy: Archives, National School of Drama, New Delhi.

Photo 1

Act I : *Nandi Path* (Benediction) is conventional *purvaranga* (preliminary) in a Sanskrit drama.

May the meditation of Siva, concentrated on Brahma owing to the intense absorption produced by the absence of any perception, protect you!—(of Siva), whose knees are wrapped round with the doubly-tightened fold of the snake, while he sits having formed a knot with his legs in the *Paryanka* posture: the operation of whose sense is so suspended as to cause the cessation of all perception, by reason of the vital airs being confined within; and who, with the eye of truth, is visualizing himself and the Universal Soul as one.

Photo 2

Act I: Sutradhara and Nati are the narrators of Sanskrit drama.

SUTRADHARA (*to the audience; angrily*): Look, worthy gentlemen, look! At the expense of *my* food, a good husband is sought for in the next world!

NATI: Be pleased, sir, be pleased! I have observed the fast that you yourself should be my husband in the next birth.

Photo 3

Act I : Sakara is a unique character in Sanskrit drama; a combination of comic-villain not found elsewhere. His incorrect accounts of mythology and history coupled with pronouncing 's' as 'shh' stimulates *hasya rasa* (sentiment of mirth).

SAKARA: Here is Vita, there is Cheta. Vita and Cheta! Cheta and Vita! As for you, you both stand in a corner. (*again searches, and catches* Radanika *by the hair*). Friend, friend, I have this time really caught her. I *have* caught Vasantasena!—

Like Chanakya seizing Draupadi, I have caught her by her massive hair. as she was indicated by the scent of her chaplets while running away in the dark.

Photo 4

Act I : Vasantasena, a courtesan in love with Charudatta, a poor virtuous Brahmin.

CHARUDATTA: This house is not a fit place for keeping a deposit.

VASANTASENA: Sir, that is not true, for deposits are entrusted to *persons*, and not to *houses*.

Photo 5

Act II : Vasantasena confesses her love for Charudatta to attendant Madanika.

MADANIKA: I can see from Your Ladyship's absent-mindedness that, you have fallen in love with somebody whom you have treasured up in your heart.

VASANTASENA: Rightly have you guessed! Clever in fathoming the hearts of others, you are indeed Madanika.

MADANIKA: Well, I am glad, I am glad indeed! Herein you are indeed obliging the mighty Cupid, who is a glorious festive joy to all young persons. So Your Ladyship will please tell whether he is a king, or a king's favourite, who is (thus desired) to be served!

VASANTASENA: Girl, I want to *enjoy*, not to serve.

Photo 6

Act II : Gambler is a King, in his own right.

GAMESTER: How now! We have a wooden idol here!

MATHURA: No, not at all; it is a stone image. (*he rocks it in various ways; then making a sign*). Never mind! Come, we shall have a game at gambling. (*they play many sorts of games at dice*)

SAMVAHAKA (*variously restraining his longing to play the game himself; to himself*). Ah!—

The rattling sound of dice stirs the heart of a penniless person, just as the sound of drums does that of a king who has lost his kingdom! I know that I shall never again play — gambling is like taking a leap from the summit of the Meru mountain! But still the rattle of dice, sweet as the song of the Cuckoo, captivates the heart!

Photo 7: Stealing is considered as one of the sixty-four arts in Indian tradition.

Act 3 is the best example of *angika abhinaya* (physical theatre) in Sanskrit literature.

VIDUSHAKA (*talking in his sleep*): Friend, I see something like a hole! I see someone like a thief! So take hold of this casket containing gold ornaments.

SARVILAKA: What! Can it be that he, knowing that I entered here, is mocking me that he is poor! Shall I then kill him?

Or is he talking in sleep owing to lightheadedness ? (*Seeing*). Ha! This thing, wrapped up in a worn-out bathing-garment, is really a casket of gold ornaments, as it gleams in the light of the lamp. All right; I'll take it —

Or rather, it is not proper to injure a person of high birth but poor like myself. So I'll go away.

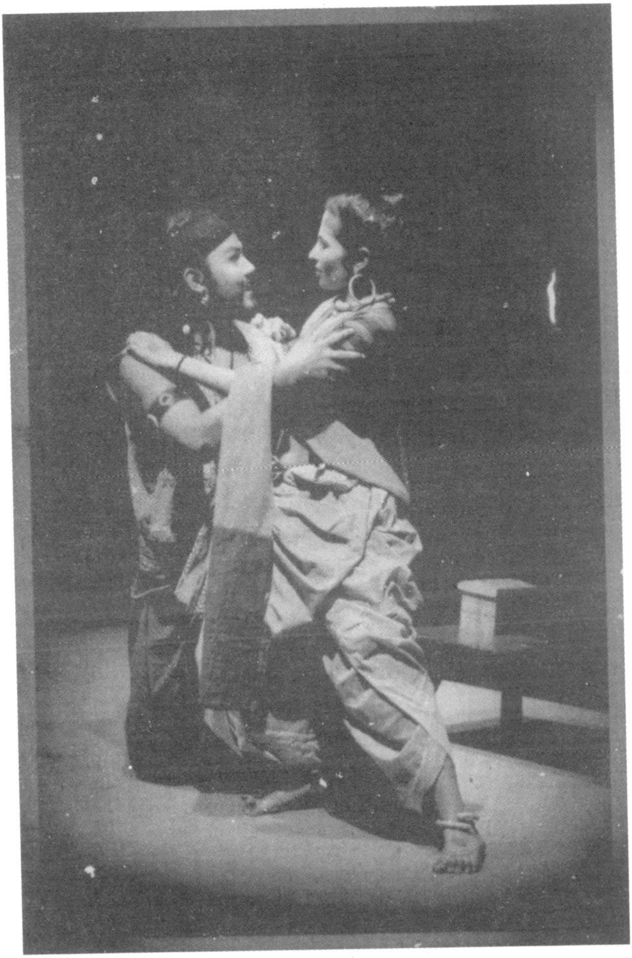

Photo 8

Act IV : Sarvilaka gets money to liberate his beloved, Madanika from Vasantasena.

MADANIKA: Sarvilaka, I did ask Her Ladyship, when she said: 'If I myself was free, I would set free all my servants from their bondage, even without any money whatever.' So, Sarvilaka, how could you ever possess so much money, so that you could liberate me from the services of Her Ladyship?

Photo 9

ACT V: Erotic sculpture poses suggesting stages of physical love.

(Vasantasena, *displaying the emotion of love, embraces* Charudatta)

CHARUDATTA (*displaying the pleasure of her touch, and embracing her in return*):

Roar still louder, O cloud! For, by your favour this body of mine, suffering the anguish of love, has become as though it were full of *Kadamba* flowers, with its hair risen on end and its passion inflamed by Vasantasena's touch.

Photo 12

ACT IX: In this act entitled, *The Trial,* Sakara puts false allegations of Vasantasena's murder on Charudatta in the court before the provost and the court scribe.

JUDGE: How! At the very commencement there is the king's brother-in-law wanting to present a case! It is just like an eclipse at sunrise, foretelling the death of some great man! Sodhanaka, the hearing of cases today will prove troublesome. Go out, my good man, and say to him that he might go away, as his suit can't be heard today.

Photo 13

ACT X: The play concludes with a *Bharatavakya*, an expression of good wishes for general prosperity; it became a convention in classical Indian drama.

May cows yield plenty of milk, the earth produce all kinds of grains in abundance, the clouds shower rain in due season, and the winds blow so as to gladden the hearts of all people! May all creatures always rejoice, the Brahmanas be ever respected, and the virtuous become rich. And may righteous kings subdue their foes and govern the earth justly!

Bibliography

Books

Banerjee, Biswanath (1999). *Shudraka*. Makers of Indian Literature. New Delhi, India: Sahitya Academy.

Bharata. *Natyashastra*. Trans. And ed. Manomohan Ghosh. 2nd rev. ed. Volume 1. Calcutta: Manisha Granthalya, 1967. Volume 2. Calcutta: Asiatic Society, 1961.

Keith, A. Barriedale. (1924). *The Sanskrit Drama*. London: Oxford University Press. 1971

Lal, Anand (2004). ed. *The Oxford Companion to Indian Theatre*. New Delhi: Oxford University Press. 2004

Nandikeswara. *Abhinayadarpanam*. Trans. Manomohan Ghosh (1975). Calcutta: Manisha Granthalya.

Oliver, Revilo Pendelton (1938). "Introduction to 'The Little Clay Cart' " in *Illinois Studies in Language and Literature* 23. Urbana: University of Illinois Press.

Pati, Madhusudan (1991). *Sanskrit Drama: Essays in Revaluation*. New Delhi: Amar Prakashan.

Prasanna. (2013). *Indian Method in Acting*. New Delhi: National School of Drama, 2014

Rangacharya, Adya (1971). *The Indian Theatre*. New Delhi: National Book Trust.

Richmond, Farley P. (1990). eds. Farley P. Richmond, Darius L. Swann, and Phillip B. Zarrilli, *Indian Theatre: Traditions of Performance*. New Delhi: Motilal Banarsidass, 2007

Sastri, Gaurinath (1960). *A Concise History of Classical Sanskrit Literature*. Calcutta: Oxford University Press. 1960

Singh, Upinder (2008). *A History of Ancient and early Medieval India: From the Stone Age to the 12th century*. New Delhi: Pearson Longman, 2008

Stoler Miller, Barbara (1993). ed. "Two classical Indian plays: Kalidasa'

Sakuntala and Sudraka's Little Clay Cart." *Masterworks of Asian literature in comparative perspective: a guide for teaching (Columbia project on Asia in the core curriculum).* M.E. Sharpe, 1994

Stoler Miller, Barbara (1984). ed. *Theater of memory: the plays of Kalidasa.* New York: Columbia University Press.

Warder, A.K. (1989). *Indian Kavya Literature.* South Asia Books, 1989

Brochures

Mrichchhakatika, Second Year Students Production. Director: Devendra Raj Ankur. National School of Drama, New Delhi. 1992

Mitti Ki Gadi, Third Year Students Production. Director: Robin Das. National School of Drama, New Delhi. 1998

Mitti Ki Gadi, Anukriti, Hindi Natya Samiti, Miranda House, University of Delhi. Director: Ramji Bali. 2009

Habib Tanvir, 1923-2009, A Photographic Exhibition. National School of Drama, New Delhi. Curator: Amal Allana. 2010

Online

Review of KS Rajendran's *Mrichchhakatika* in *The Hindu*: http://www.thehindu.com/todays-paper/tp-features/tp-sundaymagazine/an-ancient-world-comes-alive/article1436915.ece

Begum Zaidi and Hindustani Theatre: https://begumqudsiazaidi.wordpress.com/

THIRTEEN PLAYS OF BHĀSA

Translated into English by
A.C. Woolner
Lakshman Sarup

Pratijñāyaugandharāyaṇa, Svapnavāsavadattā, Cārudatta, Pāñcarātra, Madhyamavyāyoga, Pratimā-nāṭaka, Dūtavākya, Dūtaghaṭotkaca, Karṇabhāra, Urubhaṅga, Avimārka, Bālacarita, Abhiṣeka.

WORKS OF KĀLIDĀSA

(Two Volumes)
Edited by
C.R. Devadhar

Volume I comprises three famous plays of Kālidāsa : *Abhijñānaśākuntalm, Mālavikāgnimitram* and *Vikramorvaśīyam.*

Volume II contains five Kāvyas of Kālidāsa : *Śṛṅgāratilaka, Ṛtusaṁhāra, Meghadūta, Kumārasambhava* and *Raghuvaṁśa.*